The Lake District is an enchanting area of natural beauty which also has many interesting geological features. Visited by hundreds of thousands of people each year, it has much to fascinate those with an active interest in rocks, minerals and fossils. This book provides a reliable field guide to many of the most important localities.

Written by members of the Cumberland Geological Society, it covers a wide area and takes in rocks ranging in age from Ordovician (500–435 million years ago) to Pleistocene (1.8–0.01 million years ago). There are 13 self-contained itineraries which are fully illustrated with maps, diagrams and photographs. A general introduction presents the geological history of the area and a glossary of specialised terms will help those with little knowledge of the subject.

ROCKS AND FOSSILS
Editor: J. A. G. Thomas

1 *Snowdonia* M. F. Howells, B. E. Leveridge and A. J. Reedman

2 *The Weald* Wes Gibbons

3 *The Peak District* I. M. Simpson

4 *The Lake District* The Cumberland Geological Society

About the authors
Tom Shipp, General Secretary of the Cumberland Geological Society, edited this book. The contributors, all members of the society, are Kenneth W. Bond, Morley F. Burton, Frank H. Day, Mervyn B. Dodd, Fred Jones, Derek Leviston and John W. Rodgers.

Every effort has been made to ensure that the information presented in this book is accurate and up to date. However, readers are invited to draw our attention to any inaccuracies, particularly in the information concerning access to land covered in the excursions, by writing to the authors, c/o Unwin Paperbacks, PO Box 18, Park Lane, Hemel Hempstead, Herts HP2 4TE.

The Lake District

THE CUMBERLAND GEOLOGICAL
SOCIETY

London
UNWIN PAPERBACKS
Boston Sydney

First published in Unwin Paperbacks 1982

Unwin® Paperbacks,
40 Museum Street, London, WC1A 1LU, UK

Unwin Paperbacks,
Park Lane, Hemel Hempstead, Herts. HP2 4TE

George Allen & Unwin Australia Pty Ltd.,
8 Napier Street, North Sydney, NSW 2060, Australia

British Library Cataloguing in Publication Data

The Lake District. – (Rocks and fossils; 4)
1. Geology – England – Lake District (Cumbria)
I. Cumberland Geological Society
II. Series
554.27′8 QE262.L2
~~ISBN 0-04-554007-1~~

Set in 9 on 11 point Times by Typesetters (Birmingham) Limited, and printed in Great Britain
by Hazell Watson and Viney Ltd., Aylesbury, Bucks.

Foreword

Over the past few years there has been increasing public interest in the geological sciences. This derives from an increased awareness of their twofold role: first, they sketch out the history of our planet, and secondly they help to provide the mineral resources on which modern society depends. Although this spread in interest is to be welcomed, it can lead to certain undesirable consequences as more and more visitors come to examine a finite number of instructive rock exposures.

The authors and publishers of this book are fully aware of the need to make the best possible use of the outcrops it describes and have accordingly consulted with the Nature Conservancy Council, the official body responsible for conservation in Britain. All their efforts however will be nullified if today's readers choose to ignore their responsibility to students of the future.

It must be emphasised that the vast majority of geological outcrops are in private hands and that access to them is through the goodwill of the owners and occupiers. If lost, this goodwill will be difficult, if not impossible, to regain – it can take only one careless act to cause offence. Further, many geological localities can lose their interest through the cumulative effects caused by the unnecessary use of hammers. To observe and record, collecting any necessary specimens only from fallen rock, will in general give a better understanding of geology than will a physical assault on selected portions of the rock face. The indiscriminate use of hammers is thus as profitless as it is damaging.

The authors and publishers of this guide have done their best to ensure the maximum benefit from your visit to the Lake District; it remains for you to ensure that the same benefit remains available to your successors.

DR G. P. BLACK
Nature Conservancy Council

Foreword

Preface

The Cumberland Geological Society was formed in 1962 by a group of enthusiasts led by the late Charles Edmonds and by Edgar Shackleton, who continues to take an active part in its affairs. Since the inception of the Society, a monthly programme of meetings has been maintained, with lectures and films in winter and field excursions in summer.

The Society is based on the northwestern coast of Cumbria, geographically isolated from major centres of learning. Whilst most other regional geological societies may easily call upon local universities or regional offices of the Institute of Geological Sciences for assistance in the planning and manning of their activities, the Cumberland Geological Society has, by virtue of its location, been forced to adopt a measure of Cumbrian self-reliance in evolving its policies and conducting its meetings. Because of this, members, sometimes starting with little but enthusiasm for the subject, have been encouraged to plan and lead excursions, and by doing so have unearthed localities unrecorded or passed over by earlier investigators. Many of these excursions over the past 20 years have been recorded in the *Proceedings of the Cumberland Geological Society*. With this in mind, a request from Mr J. A. G. Thomas, Geological Editor of this series, led to a decision that members of the Cumberland Geological Society should write this book. Accordingly several members were approached with the result that a baker's dozen of excursions were written up by eight contributors.

We have tried to achieve a number of objectives. First, we hope that each excursion will provide sufficient information to guide the reader quickly and safely from location to location. By no means all that is of potential interest has been described in the text; the reader should be prepared to investigate, to evaluate and to interpret critically and to his own satisfaction the field evidence that he observes. The Society would welcome a feedback of observations along these lines. Second, we have tried to provide a balanced view of the geological scene in Cumbria. In a region where the rocks span the vast period of time from the Ordovician to the Pleistocene, where the present climate changes from temperate maritime to upland sub-Arctic in the space of 20 km, and where the topography includes rocky coasts, estuaries and glacially moulded mountains and lowlands, the scope for investigative fieldwork is immense. Indeed a recent review by the Nature Conservancy Council (1980) has shown that the number of student days spent by university parties alone in the Lake District has increased from 1300 in 1971–2 to 5200 in 1977–8. This was the largest increase in geological usage of all the regions surveyed in England. Only the Isle of Arran in Scotland receives a greater influx of geology students in proportion to its area. This upsurge in the use of the Lake District by university parties is without doubt amplified by school groups and interested visitors.

In this volume we have tried to guide the reader away from the over-popular high fells into localities, some well known, some less so, where aspects of most of the major episodes that have contributed to the present geological scene in Cumbria may be investigated.

Finally, we have tried, by including a glossary of terms, and a bibliography and list of references, to provide the means by which both beginners and more experienced investigators may extend their understanding of the geology of the Lake District and its environs.

T. SHIPP
Cumberland Geological Society

Contents

Foreword *page* v
Preface vii
Geological time scale x
Geological succession in the Lake District xi
List of contributors xii

PART 1: GEOLOGICAL BACKGROUND

PART 2: FIELD EXCURSIONS

1 *The Shap Granite* 15
2 *Permian rocks of the Eden Valley* 21
3 *The Mell Fell Conglomerate* 27
4 *The Armboth Dyke, Thirlmere* 35
5 *The Skiddaw Granite in Sinen Gill* 40
6 *The Carrock Fell region* 50
7 *The Carboniferous Limestone between Caldbeck and Uldale* 58
8 *Borrowdale* 65
9 *The Buttermere Valley* 74
10 *St Bees Headland* 86
11 *The Granite of lower Eskdale* 93
12 *The Appletreeworth area* 106
13 *Greenscoe Quarry: an Ordovician volcanic vent* 115

APPENDICES

I *Bibliography, references and excursion reports* 122
II *Museum facilities in Cumbria* 125
III *Glossary* 126

Index 132

GEOLOGICAL TIME SCALE

Era		*Period*		*Age to base, Ma*
CAINOZOIC		Quaternary	Holocene (Recent)	0·01
			Pleistocene	1·8
		Tertiary	Pliocene	6
			Miocene	23
			Oligocene	38
			Eocene	55
			Palaeocene	65
MESOZOIC		Cretaceous		140
		Jurassic		195
		Triassic		230
PALAEOZOIC	UPPER	Permian		280
		Carboniferous		345
		Devonian		395
	LOWER	Silurian		435
		Ordovician		500
		Cambrian		570

The three eras above may be grouped together as the Phanerozoic

Precambrian time is divided as follows:

Proterozoic	2600
Archaean	
Age of the Earth	4500

The ages given are approximations

Geological succession in the Lake District and its immediate surroundings

Stratigraphical divisions		*Principal lithological groups and formations*		
QUATERNARY	Devensian		glacial deposits	
JURASSIC	Liassic		calcareous shales	
TRIASSIC		MERCIA MUDSTONE GROUP SHERWOOD SANDSTONE GROUP	Stanwix Shales Kirklinton Sandstone St Bees Sandstone	
PERMIAN	upper		Eden Shale, St Bees Shale with evaporites	
	lower		Penrith Sandstone, Brockram	
CARBONIFEROUS	Westphalian	COAL MEASURES	upper (barren, red beds) middle lower	
	Namurian	MILLSTONE GRIT	Roosecote Mudstone Hensingham Grit First Limestone	
	Dinantian	CARBONIFEROUS LIMESTONE	Gleaston Formation Urswick Limestone Park Limestone Dalton Beds Red Hill Oölite Martin Limestone Basement Beds	2nd Limestone 3rd Limestone 4th Limestone 5th Limestone 6th Limestone 7th Limestone Cockermouth Lavas Basement Beds
DEVONIAN			Mell Fell Conglomerate	
SILURIAN	Downtonian		Scout Hill Flags	
	Ludlow		Kirkby Moor Flags Bannisdale Slates Coniston Grits Coldwell Beds	
	Wenlock		Brathay Flags	
	Llandovery		Stockdale Shales	Browgill Beds Skelgill Beds
ORDOVICIAN	Ashgill– –Caradoc	CONISTON LIMESTONE GROUP	Ashgill Shales Applethwaite Beds Stockdale Rhyolite Stile End Beds Drygill Shales	
	Caradoc– –Llandeilo	BORROWDALE VOLCANIC GROUP	Yewdale Breccia Wrengill Andesites Lincomb Tarns Ignimbrites Lickle Rhyolite Dunnerdale Tuffs Airy's Bridge and Birk Fell ignimbrites and rhyolites Ullswater and Honister Andesites	
	Llanvirn	EYCOTT GROUP	Tarn Moor Mudstones High Ireby Formation Binsey Formation	
	Arenig	SKIDDAW GROUP	Latterbarrow Sandstone Kirk Stile Slates Loweswater Flags Hope Beck Slates	

Igneous intrusive rocks, mainly associated with the Caledonian orogeny

granites	Skiddaw, Shap, Eskdale	diorite microgranite dolerite picrite minette	occurring as sills, dykes and volcanic necks
granophyres	Buttermere and Ennerdale, Carrock Fell		
microgranite	Threlkeld		
gabbro	Carrock Fell		

List of contributors

Kenneth W. Bond, BSc, FGS Formerly an oil geologist with the Burmah Oil Co. and now owner and manager of the Red House Hotel, Keswick
Excursion 8

Morley F. Burton Treasurer of the Sports Association at British Nuclear Fuels Ltd; Librarian of the Cumberland Geological Society
Excursion 4

Frank H. Day, MSc, PhD, FRSC, FGS Formerly Vice-Principal and Head of Science at Carlisle Technical College; President of the Westmorland Geological Society
Excursion 6

Mervyn B. Dodd, MA Deputy Head of Whitehaven Grammar School; Excursion Secretary of the Cumberland Geological Society
Excursion 9

Fred Jones Formerly a Senior Scientific Officer at British Nuclear Fuels Ltd
Excursion 11

Derek Leviston, BA, CEng, MIMechE Chief Inspector at Vickers Shipbuilding and Engineering Ltd, Barrow-in-Furness Shipbuilding Works
Excursions 12 and 13

John W. Rodgers, BSc Head of Geology at Queen Elizabeth Grammar School, Penrith
Excursions 1, 2 and 3

Tom Shipp, BSc, MIGeol, FGS Head of Chemistry at Workington Grammar School; part-time tutor with the Open University and the University of Newcastle-upon-Tyne; General Secretary of the Cumberland Geological Society
Excursions 5, 7 and 10,
and compilation of background material

Part 1: Geological background

The popularity of the Lake District has arisen, one suspects, largely by virtue of its scenery, which in a relatively small area (at most 3000 km^2) is as diverse as any to be found in the British Isles. Every bend in a valley, every gap in the forest, and every hilltop, high or low, brings a fresh surprise, a different scene, even to people who have been visiting or living in the district for many years and who may consider that they know it intimately.

Why is the scenery so diversified? There are the smooth sweeping lines of the mountains about Skiddaw to the north. The rugged grandeur of Scafell and its neighbours at the centre of the district excites the interest of the climber, whilst to the south there is the gentler wooded aspect of the countryside around Windermere. The geologist knows full well that the development of a landscape is brought about by a series of natural events, first by the emplacement of a variety of rocks which may then be deformed and recrystallised in a number of different ways. These altered, deformed rocks, when exposed at the surface, are then further influenced by changing temperature, reaction with air and water, the effects of vegetation and animals, and gravity. The crumbling, weathering rock material tumbles down slope, the smaller fragments being taken up by wind or water to be deposited elsewhere as sediment – the basis of the sedimentary rocks of the future. Thus the nature of the underlying rock formations, and the effects on them of all aspects of weathering and erosion, lead to the scenic features to be found in the countryside, valleys, lakes, hills smooth and craggy, cliffs, mountains and lowlands.

A look at the geological map of the region (Fig. 1) will reveal that the shapely conical hills in the northern part of the Lake District are formed of rocks of the Skiddaw Group. The craggy mountainous belt stretching from Ullswater to Wasdale is composed of the Borrowdale Volcanic Group, whilst to the south we find the more intimate low-lying fells and gentler scenery moulded from rocks of Silurian age. This entire region of Lower Palaeozoic rocks takes the form of an inlier and may be likened to a jewel of older rocks set in a mount of newer sediments. These newer sediments, comprising Carboniferous formations partly overlain by New Red Sandstone, completely encircle the Lake District and dip away from it on all sides, as if the older inlying rocks had punched their way upwards, doming and fracturing as they rose. This view of the Lake District Dome was elegantly expounded by J. E. Marr in his classic *Geology of the Lake District* published in 1916. The story has been taken a stage further by M. H. P. Bott in the Yorkshire Geological Society's collected papers, also entitled *The geology of the Lake District*, published in 1978. Professor Bott, after measuring and evaluating gravitational anomalies over northern England, suggested that the Lake District is underlain by a granitic **batholith** extending from Durham to the Isle of Man, and that this huge mass of buoyant rock is the reason why the Lakeland region has long

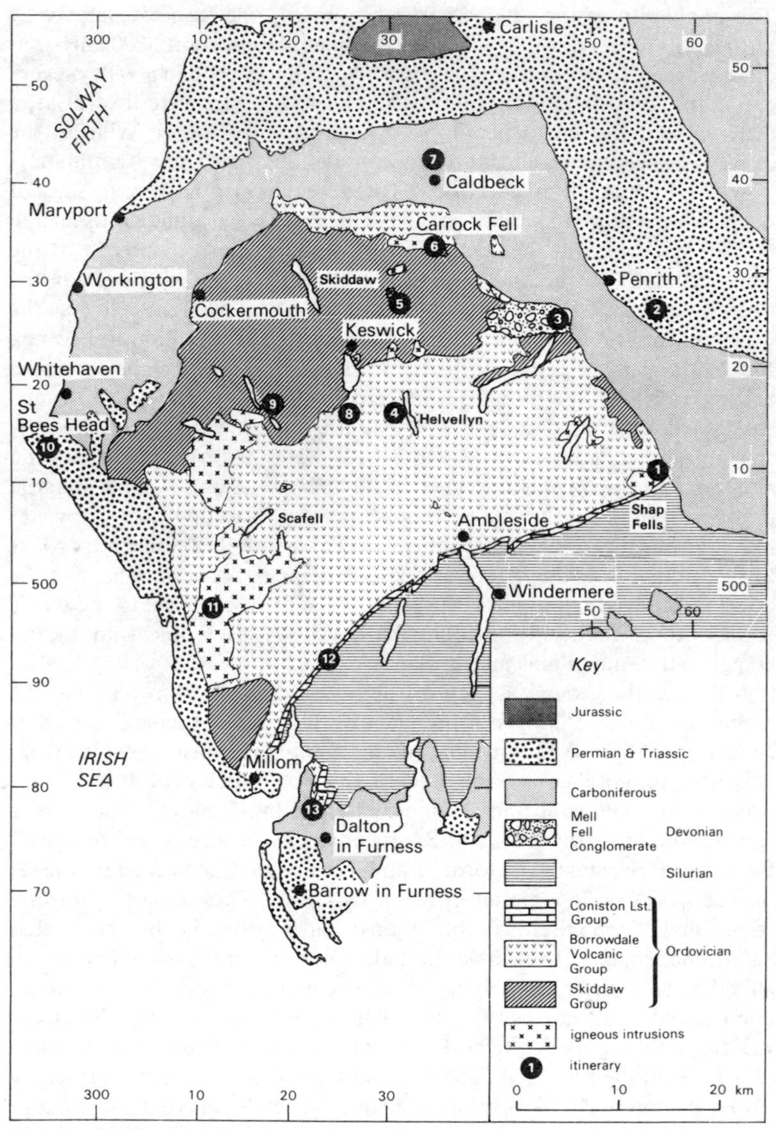

Figure 1 Geological sketch map of the Lake District showing the approximate locations of excursions 1–13. (Adapted from *British regional geology, northern England*, Plate XIII by permission of the Director of the Institute of Geological Sciences.)

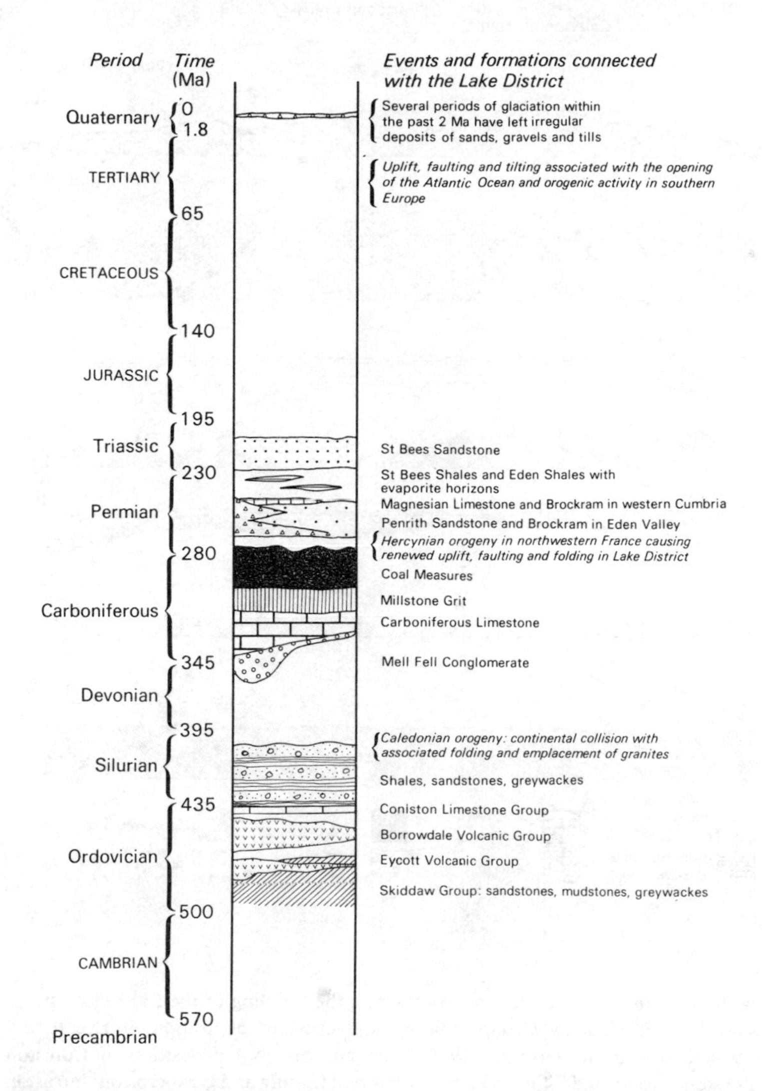

Figure 2 The geological succession in the Lake District. Geological periods unrepresented in the region are shown in small capitals. Time is shown in millions of years (Ma) before present. Numbers in circles refer to excursions described in this volume.

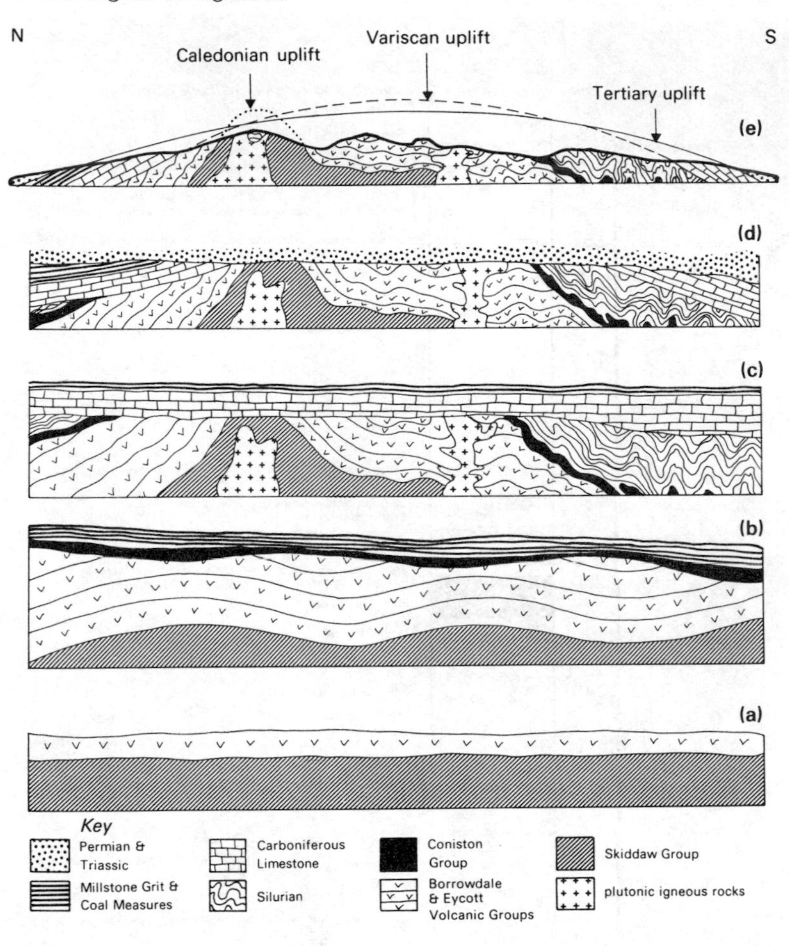

Figure 3 Diagrammatic sections to illustrate the building of the Lake District. (a) Deposition of Skiddaw Group; folding and erosion; deposition of Eycott and Borrowdale Volcanic Groups. (b) Folding and erosion; deposition of Coniston Limestone Group and Silurian Rocks. (c) Severe folding and great erosion; intrusion of plutonic igneous rocks; deposition of Carboniferous rocks. (d) Gentle folding and considerable erosion; deposition of Permian and Triassic rocks. (e) Gentle uplift, producing an elongated dome and resulting in radial drainage; erosion to present form. (Adapted from *British regional geology, northern England*, Figure 1; by permission of the Director of the Institute of Geological Sciences.)

continued to be an upland area. Professor Marr made much of the development of radial drainage patterns on a rising elongated dome of post-Triassic rocks and suggested that this has been maintained with but slight modification long after the younger rocks have disappeared from the summit of the dome. Various suggested stages in the uplift of the Lake District are shown in Figure 3.

The oldest rocks in the region, those of the Skiddaw Group, were laid down as muds and sands on the bed of an ancient ocean, called Iapetus by geologists. This once formed a wide gulf between a former continent composed of Scotland, Greenland and Labrador, and what is now England and northern Europe. Rocks of the Skiddaw Group are encountered in Excursions 3, 5, 6, 8, 9 and 13, and have clearly suffered a long and complex history of alteration and deformation since their sedimentary origins early in the Ordovician period. Due to their generally shaly nature, coupled with the fact that they are frequently folded and **cleaved**, these so-called Skiddaw Slates generally weather to form sliver-shaped fragments which easily slide away downhill and quickly become covered with vegetation. This produces scenery typified by the relatively smooth conical hills to the north and west of Keswick. True, there are crags to be found on the Skiddaw Group, but these are usually accounted for by the occurrence of sandstone units (e.g. Goat Crags, Excursion 9) or by localised hardening induced by igneous intrusion (e.g. in the **metamorphic aureole** of the Skiddaw Granite, Excursion 5).

Closure of the Iapetus Ocean during Ordovician and Silurian times led to events which have recently been neatly summarised in plate tectonic terms by F. Moseley (1977, 1978). There were two episodes of volcanic activity, the earlier producing a series of basaltic lavas, the Eycott Volcanic Group, now outcropping from Eycott Hill east of Penrith, around the north of the Skiddaw massif, westwards to Bothel village on the A595 between Cockermouth and Carlisle. These Eycott Lavas appear to be contemporaneous with the upper parts of the Skiddaw Group. The later volcanic episode produced over 4000 m of andesitic lavas, tuffs and agglomerates grading upwards into more rhyolitic lavas and **ignimbrites** (welded tuffs). Volcanic eruptions appear to have burst through the sea bed, producing a series of volcanic islands, possibly part of an island arc system similar to those to be found today in the Antilles or the Aleutians. This major volcanic outburst formed rocks of the Borrowdale Volcanic Group (Excursions 1, 3, 4, 8, 9 & 13) and in view of the colossal amount of volcanic rock to be seen it is perhaps surprising that so few actual volcanic vents have been recognised. Two have been described: Castlehead (Excursion 8) and Greenscoe (Excursion 13). The rocks of the Borrowdale Volcanic Group give us the rugged mountain scenery of the central portion of the Lake District and provide with their crags and gullies not only the highest mountain but also the finest rock-climbing terrain in England.

Towards the end of the Ordovician period the volcanic activity in the region largely died away and the sea invaded the volcanic landscape to form a thin deposit of lime-rich fossiliferous mudstones known as the Coniston Limestone Group (Excursions 1, 12 & 13). The Coniston Limestone extends across the region from the Millom area in the south-west through Coniston to Shap in the north-east. It forms a geologically important marker horizon dividing the harsh crags of the Borrowdale Volcanics to the north from the more subdued topography of the Silurian rocks to the south. These Silurian rocks originated in much the same manner as the Skiddaw Group, as muddy and gritty sediments laid down on a sea floor. However, they weather more readily than the Skiddaw rocks and form better soils; hence the softly rounded fells, the woodlands and the pastoral aspects of the southern Lake District as contrasted with the north where the Skiddaw tract supports bog, poor grassland and coniferous forest.

At the end of this long period of sedimentation came an event which, more than any other, shaped the destiny of the region. A southeastward-moving continental mass (composed of northwestern Scotland, Greenland and Labrador) came into collision with a European continental mass. Although they move very slowly, continents take a great deal of stopping, and the effects of 'the irresistible force striking the immovable object' were to alter the rocks involved very profoundly. The region was arched up into a huge ridge or anticline (Fig. 3). The more easily deformed mudstones of the Lower Ordovician and Silurian were wrinkled into tight folds (Figs 5 & 31) and squeezed like so much putty. Reorientation of platy minerals in mudstones and tuffs has produced a new rock, slate, which cleaves into thin sheets. The green slates of Borrowdale (Excursion 8) and Honister (Excursion 9) have been extensively quarried for building, roofing and decorative stone. This period of continental collision and mountain building occurring around 400 **Ma** ago is known as the **Caledonian orogeny**, as its effects are most notably seen in the Highlands of Scotland (Caledonia).

It was during this period of mountain building that great masses of granitic magma invaded the roots of the mountain ranges and crystallised under the Lake District to form a batholith, the upper portions of which are exposed today as the Shap Granite (Excursion 1), the Skiddaw Granite (Excursions 5 & 6), the Buttermere and Ennerdale Granophyre (Excursion 9), and the Eskdale Granite (Excursion 11). Associated with these granites are later-formed mineral veins containing lead, zinc, copper, iron and tungsten minerals. Of the dozens of metal mines formerly operating in the region, only one (the Carrock Mine, Excursion 6) remains operative, producing tungsten concentrates. The lead, zinc and copper mines once worked by German miners in the days of Elizabeth I are all derelict

(Excursion 5, Figs 17 & 18) whilst the last haematite mine in West Cumbria closed in 1980.

The Caledonian orogeny raised chains of mountains, possibly of Himalayan proportions, across northern Britain which during the Devonian period is thought to have lain some way south of the Equator. The effects of climate at these latitudes, coupled with the fact that land plants were few and primitive and therefore did not bind together the breakdown products of rocks, combined to erode the newly formed mountains very rapidly indeed. There is field evidence that the Shap Granite, which was emplaced in the early Devonian at a depth of several kilometres within the crust, was exposed at the surface by early Carboniferous times (Excursion 1, location **6**). Some of the products of the rapid erosion of the Devonian mountains are to be seen as the Mell Fell Conglomerate (Excursion 3), and certainly by Lower Carboniferous times the harsh Devonian mountains had been worn down to a rocky landscape of low relief over which the tropical seas advanced and retreated, producing the characteristic **cyclothems**: alternations of fossiliferous limestones, shales and sandstones so typical of the Carboniferous Limestone (Excursion 7).

The arrival of the British region at the Equator coincided with the rapid evolution of land plants, including giant versions of the present-day lowly horsetails, growing on a series of flat-lying swampy deltas extending from central Europe to eastern USA. These early tropical swamp forests produced the Carboniferous Coal Measures, now outcropping in northwestern Cumbria and extending westwards under the Irish Sea. Only one deep mine, the Haig Colliery at Whitehaven (Excursion 10), remains in operation, but there are several sites between Whitehaven and Maryport where coal is being won by opencast operations.

Towards the end of the Carboniferous further folding took place in the region, re-elevating the Lake District massif, downwarping the Eden Valley and activating the gigantic Pennine Fault system. These movements possibly arose as distant ripples of the **Variscan orogeny**, centred in northwestern France. During Permian times the Lake District massif and the Pennine ridge to the east lay in the path of northeasterly trade winds in latitudes approximating to those of the present Sahara desert. The Permian rocks show every evidence of formation in an arid windy environment (Excursion 2), with the shore of a highly saline sea lying in the vicinity of St Bees Head (Excursion 10, location **3**). To envisage the physical conditions of the Permian in Cumbria, one might at present make a tour of the Arabian Peninsula, noting the processes of erosion, dune migration and evaporation of saline waters taking place in that arid region. The succeeding Triassic rocks in Cumbria are predominantly water-laid false-bedded red sandstones full of interesting sedimentary structures but tantalisingly unfossiliferous (Excursion 10).

From the Triassic until the Quaternary the geological history of the region is extremely conjectural. A basin of Lower Jurassic rocks occurs in the Wigton–Carlisle area, but the nearest Cretaceous rocks lie on the far side of the Irish Sea. Did Jurassic and Cretaceous seas once cover the Lake District and mantle the area with oölite and soft white chalk? Professor Marr thought so, suggesting that the radial drainage was initiated on a dome-like cover of relatively uniform soft rocks now eroded completely away.

Whatever the initial cause, the river pattern in the region is still markedly radial and it does not seem to be well adjusted to variations in the underlying rock formations, though some of the major valleys do lie along faults (e.g. Thirlmere, Excursion 4) or anticlinal axes (e.g. Ullswater, Excursion 3). Attempts have been made to explain away the radial pattern by invoking structural guidance rather than initiation on a rising dome, but the problem may never be resolved.

Of the past 2 Ma or so, and certainly of the past 20 000 years, much more is known with certainty. The region has been subjected to successive glacial and interglacial episodes. How many there were is proving to be one of the geological debates of the decade. Suffice it to say that one of the main attractions of the Lake District, to geological and non-geological visitors alike, lies in the profusion and freshness of its glacial landforms. The non-discerning visitor enjoys the lakes, the craggy hillsides and the abundance of boulders. In his mind's eye the geologist can see snow avalanching on to the glacier surface in the cirque at Bowscale (Excursion 6, Fig. 20), visualise ice laden with stones grinding the bedrock at Red Brow (Excursion 8, Fig. 29), or watch the progress of the glacier snout protruding from the Stonethwaite valley at Rosthwaite (Excursion 8, Fig. 27). In imagination one can stand close to the edge of the ice sheet near Brankenwall (Excursion 11, location **2**) and hear the meltwater thundering southwards to reach open sea far to the south of an Irish Sea basin, brim full – as far as the eye can see – with glacier ice moving southwards from its source on the Scottish hills.

Following the geological excitements of former ages, the massive volcanic outbursts, the uplift of huge mountain ranges, the invasions of tropical seas and growth of luxuriant steamy forests, the aridity of deserts, and the iron grip of glaciers and their watery relaxation, present-day geological processes in the region do seem rather tame. But geological change *is* taking place. Rocks continue to be weathered. Scree material occasionally cascades down slope, often aided by humans and their animals. Soil continues to creep down hill to be whisked away by streams. The lakes are shrinking inexorably, filling up with debris, most of it naturally deposited, but some of it thrown in by small boys and their fathers on holiday. On the positive side the region is very slowly rising: there are raised beaches on the Cumbrian coast. Each of us has his own views on the way in which the climate is changing, and each has his own fears about the possible effects of man's activities on both landscape and climate.

To conclude, the geological record of the Lake District is incomplete in two important senses. First, the more information we obtain by careful fieldwork, applying increasingly refined techniques and thereby gaining a greater insight into the long and complex geological history of the region, the more problems seem to present themselves for solution. Second, the processes of geological change continue; they cannot be arrested; and indeed patient investigations into present-day processes have led geologists to a greater awareness of what happened long, long ago. The present is the key to the past, and those who would use this key must try to use all their senses, all their intellect in the field. Where better to apply this precept than in the Lake District?

Since the geology of the Lake District is complex, further details of the geological setting are given at the beginning of each excursion where needed.

A brief explanation of words in **bold-face type** will be found in the glossary. Excursion localities are also printed in bold-face type, e.g. **(2)**.

Readers who are beginning geology – and others – may find the following books helpful:

Hamilton, W. R., A. R. Woolley and A. C. Bishop. *The Hamlyn guide to minerals, rocks and fossils*. London: Hamlyn. This book contains coloured illustrations and descriptions of many of the minerals, rocks and fossils likely to be encountered.

British Palaeozoic fossils. London: British Museum (Natural History). This book contains a further selection of fossil drawings.

Read, H. H. (Ed.) 1970. *Rutley's Elements of mineralogy*. London: George Allen & Unwin. This book provides a means of identifying a wide range of minerals.

Part 2: Field excursions

1 The Shap Granite

(1 day, or part of ½ day)

Purpose: Examination of the Shap Granite and metasomatism in surrounding rocks; the unconformity of Shap Wells and its use in dating the granite; folding in the Silurian strata on Shap Fell.

Transport will be required for movement from one location to another in reasonable time. Since the countryside hereabouts is open moorland, fellwalking gear and strong footwear should be worn. Safety helmets and eye protection should be worn when visiting the quarries.

Permission is needed to visit the Blue Rock Quarry. Write, giving details of the proposed visit, and enclosing a stamped addressed envelope for reply, to Thomas W. Ward (Roadstone) Ltd, Shap Granite Co., Shap, Penrith, Cumbria CA10 3QQ. Party leaders are requested to report to the main office at the Shap Granite Works and sign the usual indemnity forms. The party must report to the foreman in charge at either quarry, so that visits are restricted to the working week.

OS maps: 1:25 000 Outdoor Leisure Map. The English Lakes SE Sheet **(1–5)**
Sheet NY 50 **(6)**
1:63 360 Tourist Map of the Lake District **(1–5)**
Only the 1:25 000 map shows all footpaths and place names.

The locations are adjacent to the A6 south of Shap village, and there are adequate parking facilities close to each. There are toilets and cafés in Shap village, and the Shap Wells Hotel **(6)** provides useful amenities. Locations are shown on the sketch map (Fig. 4).

EXCURSION DETAILS

Location 1: Folded Silurian strata, Shap Fell (555055) (Fig. 5).

A few hundred metres south of Shap Fell summit on the A6 is a fine section of strongly folded Silurian rocks. There are suitable lay-bys on either side

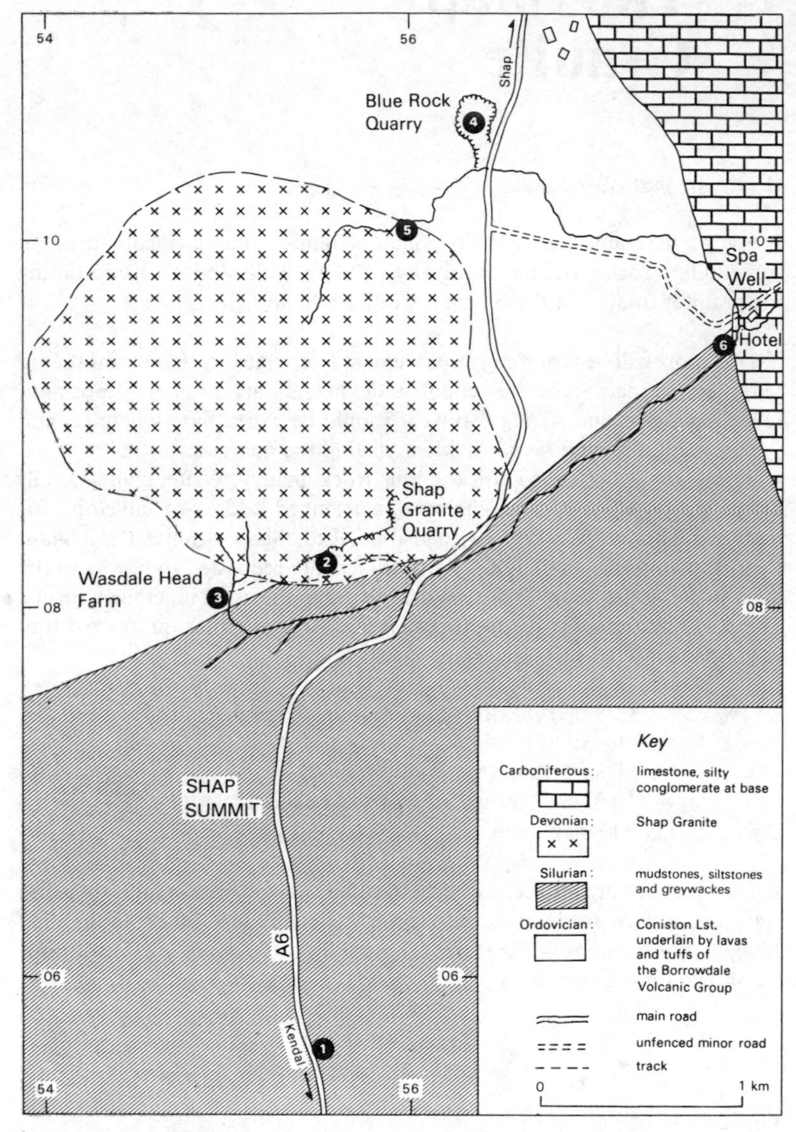

Figure 4 The Shap Granite.

Figure 5 The Shap Granite. Folded Silurian strata in the roadside at Shap Fell (555055). (Photograph: T. Shipp.)

of the road at or near the summit, and wide footpaths give good access to the exposures.

The sediments themselves belong to a group transitional between the Coniston Grit and the Bannisdale Slates, divisions of the Silurian (see inside back cover). They consist of well cleaved mudstones (laminated in places), siltstones which are less well cleaved, and quite thick greywacke beds which have very little **cleavage** at all. The sediments in some parts of the section dip steeply and uniformly to the south, but elsewhere they are very clearly folded into a series of tight, mainly asymmetrical, anticlines and synclines. This is a most useful section for detailed study of minor folds, and their related structures, such as cleavage, **joints**, **tension gashes**, and so on; and on some of the undersides of the beds there are interesting **sole markings**, mainly **flute casts**.

Location 2: Shap Granite (555084)

This well known intrusion takes the form of a **boss**, just under 8 km^2 at the surface, but believed to be a small elevated part of an extensive granitic batholith that underlies much of the Lake District. The quarry affords access to excellent exposures, but particular care should be taken to keep well away from working machinery and unstable slopes. Climbing on any of the exposed faces or slopes is extremely dangerous.

The main features of the granite itself are its characteristic **porphyritic texture**, the pink **orthoclase phenocrysts** commonly exceeding 3 cm in length, and its division into light and dark varieties. This colour difference seems to be related to the master joints, the redder, darker zones possibly being due to iron enrichment. If time (and work in the quarry) permits, variation in the proportion of phenocrysts can be studied.

Other interesting features to be seen in the quarry include mineralisation along joint planes. Pyrite and **chalcopyrite** are very common; quartz, **barite** and **fluorite** also occur; and it may be possible to find the soft, scaly **molybdenite** which seems to be rather rare now. At the western end of the quarry there is a zone of weathering that has altered the granite to a greenish crumbly material. Other such zones can be located around faults, one of which can be clearly seen just to the right of the central unworked portion.

Finally, but of considerable importance, is the existence of dark **xenoliths** throughout the granite, most of which contain pink orthoclase crystals (Fig. 6). These xenoliths have been considered either as portions of a more basic, deeper intrusion or as blocks of the surrounding Borrowdale Volcanics which were incompletely assimilated in the magma.

Figure 6 The Shap Granite. Texture of the granite showing orthoclase feldspar phenocrysts developed both in the granite and in the prominent dark oval xenolith. (Photograph: T. Shipp.)

Location 3: Wasdale Head Farm (549081). **Metamorphic aureole**

A track runs westwards from the lower part of the granite quarry to the old, ruined buildings of Wasdale Head Farm. About 100 m beyond the farm is a series of small crags above the path. The lower exposures – greyish-white granular rocks – are thermally **metamorphosed** Coniston Limestone sediments. Immediately above are grey and pink flow-banded rhyolites which can be shown to dip below the limestone group. Before returning to the road, the granite contact can be traced just north of Wasdale Head Farm.

Location 4: Shap Blue Rock Quarry (565105). **Metamorphic aureole**

This large quarry has been excavated into dark volcanic rocks of the Borrowdale Group. Their original nature has been partly obscured by both thermal metamorphism and **metasomatism** effected by the granite intrusion which lies immediately to the south. The most notable feature is the abundance of mineral veins, which are believed to have formed by the introduction of fluids from the granite along fissures in the rocks of the aureole. There is evidence that some mineralisation may have preceded the granite, and that some of the minerals present were formed by thermal metamorphism of pre-existing veins, and not by metasomatism associated with the granite. For example, a sample of **wollastonite** found in the quarry suggests the metamorphism of quartz–calcite veins, and such veins have been located in the surrounding area. Among the more common minerals are quartz, pyrite, red-brown **garnets** (usually associated with the quartz), yellow-green **epidote**, calcite and **haematite**.

Much of the original rock appears to have been a basic vesicular **andesite**. Alteration has resulted in many of the former vesicles becoming oval masses of dark green feathery **amphiboles**. Elsewhere, quartz and calcite had filled in the vesicles prior to the intrusion of the granite, and these **amygdales** are still clearly visible, though recrystallised. The whole mass of volcanic rock is thought to have been partially weathered before metamorphism, which then considerably hardened it, a property which has led to its extensive quarrying as a roadstone.

Location 5: Granite contact (562101)

The intrusive nature of the granite can be well seen at a point south of the Blue Rock Quarry. From the quarry entrance, take the track that goes to the left, alongside the plantation. Follow this track for about half a kilometre to a bridge by a gate. On either side of the bridge, in the stream bed, can be seen tongues and veins of pink granite penetrating the country rock. The latter has been extensively recrystallised to **hornfels**, but strings of former amygdales indicate that it was similar to the basic andesite found in the quarry. It is also possible to find dark xenoliths of this country rock in some of the granite, and there are small dykes of fine grained pink **aplite** in one or two places.

Location 6: Shap Wells unconformity (578095)

From the Blue Rock Quarry, take the A6 south and turn off for the Shap Wells Hotel a few hundred metres down the road. Vehicles may be parked, after first asking for permission at the hotel for parking and access, in the large car park of the hotel forecourt. The section to be examined here is in the bed and bank of the stream that flows from the south-west and it can be clearly seen on the right as the hotel is approached.

Please do not use hammers in this location.

First go upstream to the waterfall and note dark faintly laminated siltstones with occasional pyrite nodules. These beds follow the regional strike of the Palaeozoic formations (roughly NE–SW) and dip quite steeply to the south-east. They are early Silurian in age.

Immediately below the waterfall is the **unconformity**. The Silurian strata are overlain by a thin layer of friable greyish-green silty material which dips gently downstream. This, in turn, is overlain by a conglomeratic deposit which further downstream becomes almost horizontal and a pinkish-brown colour. Examination of this conglomerate will quickly reveal the distinctive orthoclase crystals of the Shap Granite, and it follows that this material could only have accumulated after the granite had been intruded, exposed and partially eroded. The conglomerate is considered to be part of the Basement Beds of the Carboniferous, so on geological field evidence the Shap Granite can be considered a Devonian intrusion. This has been neatly confirmed by isotopic dating which suggests an age of 393 Ma.

If there is time, unmetamorphosed Coniston Limestone rocks can be found, with some care, by the Spa Well just to the north of the hotel.

2 Permian rocks of the Eden Valley

(1 day)

Purpose: To study sedimentary rocks of Permian age showing the varied nature of Permian environments.

The localities chosen are some distance apart, and transport is essential. Visits to one or more of these localities may be combined with field work in adjacent areas. All quarries mentioned are disused, and thick vegetation could cause problems in summer and early autumn. Permission where required is indicated in the text (**2** and **3**). As this part of the Eden Valley is all enclosed farmland or afforested, great care should be taken not to damage gates, walls, fences, trees and crops, and to leave no litter.

OS maps: 1:50 000 Sheets 86, 90 and 91
IGS maps: 1:50 000 Sheets 24 (Penrith) and 31 (Brough)
(solid *or* drift editions)

Suggestions for parking are given for each locality. There are toilet and refreshment facilities in Penrith, Appleby and Brough.

GEOLOGICAL SETTING

The Vale of Eden lies on the northeastern edge of the Lake District and contains a broad tract of Permo-Triassic rocks. Rocks of this age are often referred to as New Red Sandstone, the term being particularly appropriate in this region. Here sandstone is the dominant rock type exposed, its warm reddish-brown colour being reflected in the houses, walls and soils. Much of the valley is covered by drift deposits, some of which are commercially exploited, for instance, the sand and gravel deposits around Lazonby, while others such as the drumlin field between Appleby and Brough, and the meltwater deposits of the Cross Fell Inlier, are of considerable topographic interest.

The purpose of this itinerary is to give a brief guide to selected exposures along the western side of the Eden Valley to illustrate the nature of the

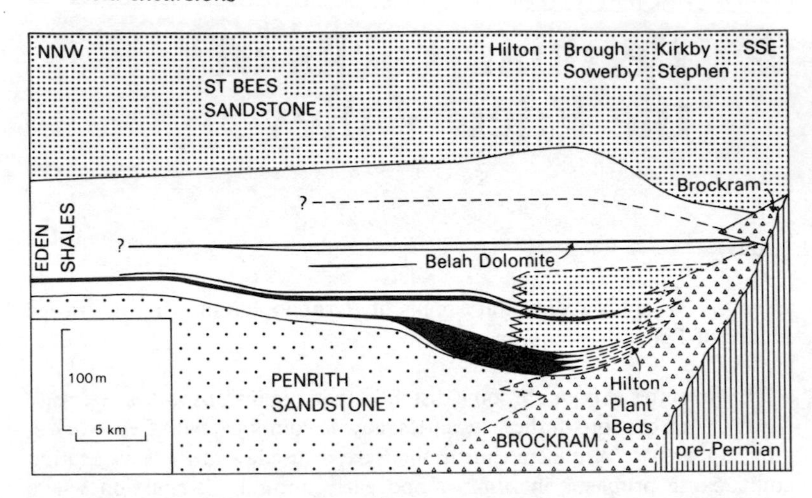

Figure 7 Vale of Eden. Section showing the interrelationships of named Permian rock division. The gypsum–anhydrite horizons are shown in black. (Adapted from *The geology of the Lake District*, Figure 65, with permission from the Yorkshire Geological Society.)

Permian rocks and indicate their origin. Two points worth keeping in mind when studying this area are as follows:

(a) At the end of the previous period, the Carboniferous, most of Britain, including northwestern England, had become part of a continental land mass as a result of the Variscan orogeny.
(b) Britain at this time lay a few degrees north of the Equator.

The three main Permian rock divisions in this area are Penrith Sandstone, **Brockram** and Eden Shales. The relationship between these divisions is shown in Figure 7.

Exposures at the following localities illustrate these divisions. With the exception of **7** they are arranged in order from north-west to south-east.

EXCURSION DETAILS

Location 1: Cowrake Quarry, Penrith (540309). Penrith Sandstone

The town of Penrith is dominated by a wooded hillside, at the summit of which stands Penrith Beacon. This prominent slope is part of a southwest-facing scarp composed of the relatively hard Penrith Sandstone. There are

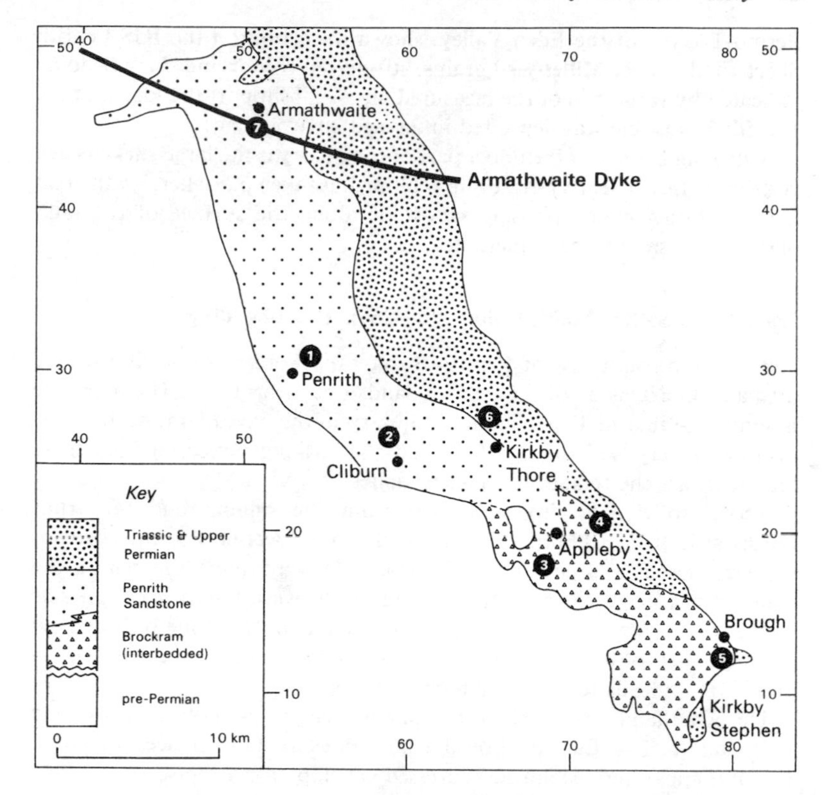

Figure 8 Post-Carboniferous rocks of the Vale of Eden.

numerous old quarries in the district, Cowrake Quarry being situated a little way to the south of the summit of the Beacon. From the town centre take any of the roads which lead up to Beacon Edge, then turn right at the top. Follow the road eastwards for about 1 km from the edge of the town. At the second bend in the road, about 300 m past the minor road on the right, is an iron gate on the left, beyond which is the track to the quarry. There is a little parking space here at the side of the road, but not enough for a coach. A small path climbs steeply off to the left just around the first bend in the track, and this leads to the main high-level part of the quarry.

The sandstone is variable in grain size, reddish-brown and composed almost entirely of well rounded and 'frosted' quartz grains with a thin coat of haematite. In some places it is very friable but in others it is very hard due to secondary precipitation of silica cement. The present regional dip is largely the result of folding which produced an asymmetrical syncline in the

Permo-Triassic of the Eden Valley, shown in section on the IGS Penrith Sheet cited above. **Millet-seed grains,** strong red colour and dune bedding indicated by variations of the measured dip – all support the idea that the Penrith Sandstone was deposited in an arid environment.

A striking feature of the upper part of the quarry is the large **slickensided** vertical surface. Clearly, strike-slip faulting has taken place here, with great local friction metamorphosing sandstone along the surface of the fault plane into a smooth pale quartzite.

Location 2: Salter Wood, Cliburn (584264). **Dune-bedding**

About 7 km south-east of Penrith is Whinfell Forest, a slightly elevated area and an extension of the Penrith Sandstone escarpment. The rock here is similar to that in **1**, but in an old quarry in the wood large-scale cross-bedding is very well displayed and gives a striking indication that these sediments are the remains of ancient dunes.

Access to this locality is gained from the minor road that runs northwards from Cliburn (587249) to the A66. About 1.5 km north of Cliburn there is a large bend in the road. Halfway round the bend is the gated entrance to a forest track that leads to the west. On the opposite side of the road is a wooden fence across an opening in the stone wall. An old path runs off to the left from this opening, sub-parallel to the wall, and in about 100 m leads to the old quarry sections.

This woodland is privately owned and any enquiries about access should be made to The Earl of Lonsdale's Estates, Estate Office, Lowther, Penrith, enclosing a stamped addressed envelope for a reply.

Location 3: Burrells (676180). **Brockram**

While desert sands were accumulating as dunes in the lower parts of the Permian sedimentary basin that covered this region, nearer high ground, and particularly in the area that is now the south-east part of the Eden Valley, coarse gravels were being deposited. These gravels are now seen as breccias, locally known as brockram, and they consist principally of irregularly shaped large **clasts** of Carboniferous Limestone, often **dolomitised,** set in a reddish sand matrix. Clasts of sandstone can also be found. These deposits are thought to have formed as screes or **wadi** wash-out deposits, the material being derived from the then newly uplifted Carboniferous strata to the east.

An excellent section of brockram can be seen at Burrells, about 2 km south of Appleby on the B6260. The relatively hard breccia forms a low but distinct escarpment on either side of the road. Opposite a cottage called the Gate House is a field containing an old quarry with a steep wall of brockram. *Please do not hammer.* Access to this is via a gate from the road, but parking may be difficult as the road here is narrow and winding. A large vehicle would have to be left at Hoff about 500 m down the road and, for a large party, permission should be sought from Friendship Farm, Burrells.

Location 4: Hilton Beck (720207). Eden Shales

Following the formation of aeolian sediments of the Penrith Sandstone came a period when water deposition became more dominant. The earliest of these deposits, the so-called Eden Shales, are to be found in the Hilton area quite well exposed in a stream section that runs parallel to the road just to the west of Hilton Village. The pleasant valley is one of numerous glacial **meltwater channels** that exist in this area, and while this one acts as the present-day drainage line for Hilton Beck, many of the others are, for most of the time, dry valleys.

The best section is to be seen between the two footbridges (716204 & 720207); please keep to the footpath and *do not hammer.* It consists of alternating beds of buff sandstones and siltstones, with fragments of fossil plants occurring abundantly in the finer-grained sediments, especially within 100 m of the upstream bridge. **Evaporite** horizons are an important feature of these beds but they do not crop out at the surface, having been dissolved away. About 400 m above the bridge (723205) is a hard band of **dolomite**, the Belah Dolomite, which is believed to represent a marine incursion into an area which had been largely salt flats and coastal plain. A more complete sequence of these beds has been obtained from the Hilton borehole and is to be found in the IGS publication, *The geology of the Cross Fell area.*

Location 5: Brough Sowerby (795121). Faulted Penrith Sandstone and Brockram

Return to Appleby and take the A66 to Brough. There is a good section of dune-bedded sandstone on the left-hand side of the road at the bottom of the hill in the centre of Appleby. The main road runs close to the Pennine Fault line between Warcop and Brough, and from the road immediately to the west of Brough a large anticline in the Carboniferous Limestone can be clearly seen in the fellside around Helbeck.

Turn right in the middle of Brough and travel south along the A685 for about 3 km until reaching a large lay-by on the left (eastern) side of the road. At this point the River Belah runs beneath the road.

About 300 m upstream is a cliff section in which are displayed alternating bands of brockram and soft red sandstone. This interbanding allows two small normal faults to stand out very clearly in the main body of the cliff, and also shows how one kind of deposit alternated with another. *Please do not hammer.*

Location 6: Evaporite (gypsum and anhydrite) deposits

Another feature of the Permian rocks of the Eden Valley worth mentioning is the existence of calcium sulphate evaporite deposits within the Permian strata, deposits which are mined as gypsum near Kirkby Thore (646267), but which exist as **anhydrite** further north, for example, in the now disused Long Meg mine (563378) near Langwathby. Such sediments complete the picture of a hot semi-arid subsiding basin with an occasional marine transgression either flooding the region or just raising the water table enough to form saline lakes in topographic hollows.

Location 7: The Armathwaite Dyke

This feature, well worth a visit, is not of Permian age but Tertiary. Towards the northern end of the Eden Valley, the Penrith Sandstone is intruded by a thick **dolerite dyke**. This crosses the River Eden just south of the village of Armathwaite, and forms a fine set of rapids (504453). The best approach is via the path that runs along the east bank of the river from Armathwaite Bridge (the left bank going upstream). After about 500 m descend to the dyke, which is clearly visible as a series of **jointed** blocks right across the river, by way of a minor path. Although the outcrop of the dyke is discontinuous across the country, its trend and outcrop pattern link it to the Cleveland Dyke of the North Yorkshire Moors.

3 The Mell Fell Conglomerate

(1 day or ½ day)

Purpose: Examination of the Mell Fell Conglomerate (**1,2,4,8**) and its relationship to underlying formations (**3,5,6,7**). The conglomerate occupies an area of about 8 km × 3 km to the north-west of Ullswater, forming the rounded hills of Great Mell Fell and Little Mell Fell to the west, with Soulby Fell and Dunmallard to the east. Its outcrop may be mapped on a suitable base map by a group using information from the various exposures and topographic features.

OS maps: 1:25 000 Outdoor Leisure Map, The English Lakes, North-East sheet. (This shows all the detail required.)
1:50 000 Sheet 90
1:63 360 Lake District Tourist Map

The locations to be visited lie on, or adjacent to, roads or footpaths. The excursion starts at Pooley Bridge (472244), where there are shops, toilets, a car park and an information centre. It should not prove difficult to park a car or minibus along minor roads, but coaches are advised to keep to the A-roads.

GEOLOGICAL SETTING

Though its base is not exposed the conglomerate is inferred to rest unconformably on an irregular erosion surface composed of rocks of the Skiddaw Group, which tend to form the lower-lying, mainly drift-covered areas around the lake, and of the Borrowdale Volcanic Group (BVG), which give the craggy hills and ridges. These two formations crop out around Ullswater in a rather complex fashion because of extensive faulting, but they can be quite easily distinguished by their topography. To the north of the conglomerate is the escarpment of the overlying Carboniferous Limestone, which in this area runs approximately east–west.

Old Geological Survey maps show the conglomerate as part of the Carboniferous Basement Beds. However, despite the absence of any useful fossil material, its coarse texture, its poor sorting and strongly red-stained

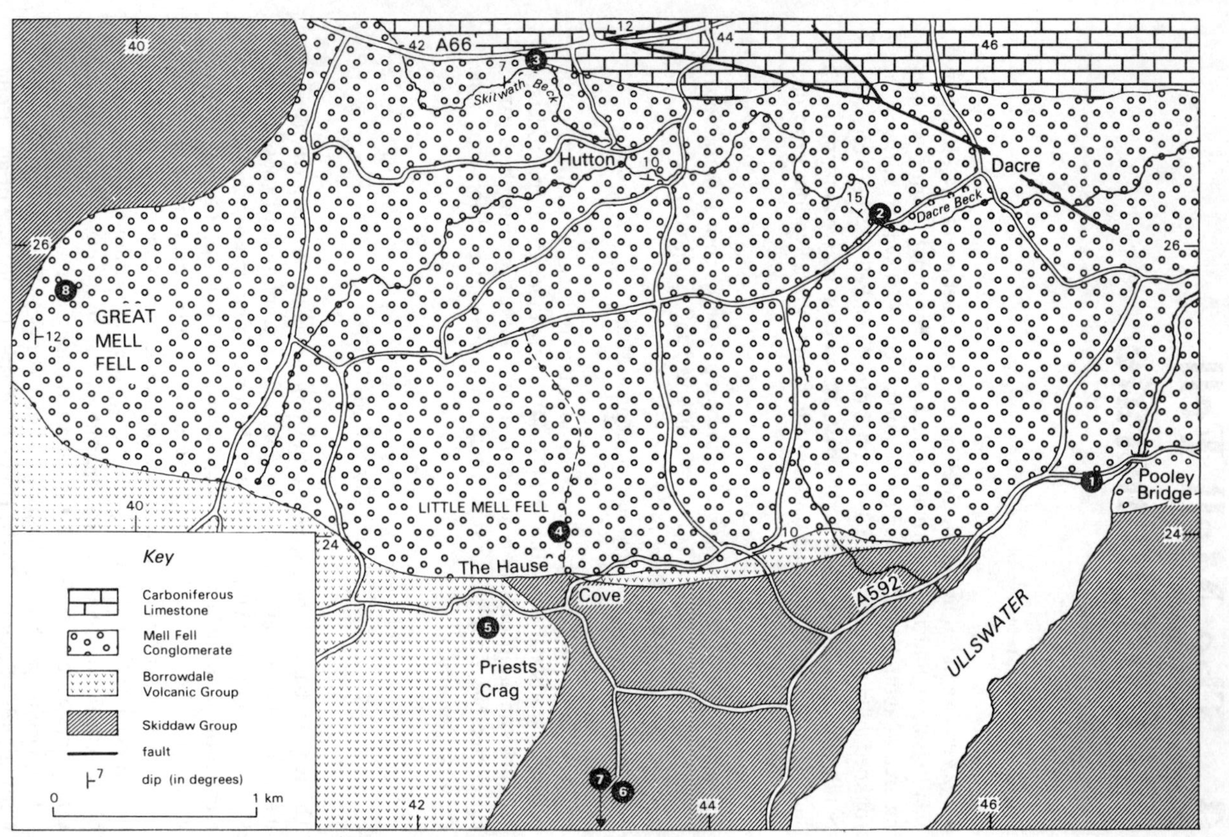

Figure 9 Sketch map of the Mell Fell Conglomerate north-west of Ullswater.

character, together with a thickness of at least 300 m, suggest that it represents a continental torrential deposit laid down in an arid or semi-arid environment. This in turn makes it most likely to be a remnant of Old Red Sandstone material, produced by rapid erosion of the newly formed Caledonide mountains in Devonian times.

EXCURSION DETAILS

Location 1: Pooley Bridge (472244). **Coarse conglomerate**

Good exposures of typical conglomerate can be seen in the bedrock of the lake shore or in sections on the opposite side of the road. From Pooley Bridge, where there are adequate parking facilities, follow the main road to Patterdale in a westerly direction. Just south of the landing stage, about 300 m from the bridge (467243), it is possible to get onto the rocky foreshore of Ullswater over a low wall.

The rock here is a very coarse conglomerate, consisting of sub-angular pebbles, **cobbles** and occasional boulders, set in a reddish gritty matrix. The majority of **phenoclasts** are of greyish-green **greywacke**, very like those found in the Silurian beds in the southern part of the Lake District and on Shap Fell (Excursion 1). There are some volcanic pebbles and, in places, patches of coarsely crystalline calcite, possibly derived from limestone pebbles which are known to occur in the conglomerate. Bedding is virtually non-existent. The shoreline exposures are useful for examining the smooth water-worn surfaces of the rock, while across the road there is a short vertical section. Alternatively, a much larger section (**1a**) can be examined further along the road. Go past the road junction to the small boat-house on the lake side, opposite which is a lay-by. A section of the conglomerate can be safely examined between the lay-by and Flosh Gate (459239). A notable feature of this deposit is the size of the boulders, some of which approach 1 m in diameter.

Location 2: Dacre End (452261). **Bedded conglomerate**

Return towards Penrith via the A592, and turn off left for the village of Dacre. Pass through the village, turning left again in the centre, and make for the bridge over Dacre Beck about 800 m west of the road junction (452261). The road is quite narrow, but small vehicles can be parked with care just before the bridge.

Well bedded conglomerate is clearly displayed in the stream bed above the bridge. The rock is finer grained than at Pooley Bridge and it dips at a low angle (about 15°) to the north-east, but again, the clasts are predominantly greywacke.

The finer-grained nature of the rocks here has been explained by considering these sediments as the peripheral deposits of a series of fan deltas, the coarser rocks further south being closer to the torrential mouth. An alternative explanation would be to consider the finer sediments as representing a later series lying above the older, coarser rocks. This second explanation depends upon accepting the dip of the formation as being tectonic in origin, caused by later earth movements; but others believe that a major component of the dip may be depositional, with gravels and sands deposited on sloping surfaces.

Location 3: Skitwath Beck, Hutton (428273). **Basal Carboniferous and underlying conglomerate**

This locality is probably the best place to see the relationship between the conglomerate and younger rocks. Nowhere is the actual junction between the conglomerate and the overlying Carboniferous exposed, but in the valley of Skitwath Beck it is possible to infer that Mell Fell type sediments are overlain by Carboniferous Limestone with its own basal conglomerate.

Vehicles can be parked in the lay-by on the A66, just west of the cross-roads (428274). The best exposures are, in fact, just a short distance from this point on the other side of the road, but there is no access across a barbed wire fence, so a longer way round is necessary. Follow the main road west for about 500 m until reaching a set of steps on the northern bank of the cutting. Immediately opposite are two gates. Go through the right-hand gate and follow the old footpath that runs alongside the wall down to the corner of the wood.

Turn left and follow the path across the fence and along the top of the steep bank. This path runs parallel to the fence line at the edge of the wood and may be quite difficult after heavy rain. It is overgrown in some places. At a point opposite the eastern end of the lay-by is a clearing in the trees that descends to the stream below. Two concrete manholes can be seen down this slope. Descend carefully to these, then move left to where a small stream has exposed bedrock in the steep bank. The rock is rather weathered and friable but is easily recognised as a yellowish fine-grained conglomerate, quite different from the Mell Fell formations. The latter can be seen on the other side of Skitwath Beck, well exposed as steep walls, and dipping gently to the north-east. Climbing back up the bank, yellowish-brown dolomitised Carboniferous Limestone can be seen towards the top. This, then, represents the northern edge of the Mell Fell Conglomerate outcrop, and from here eastwards the boundary can be inferred to run along the base of the escarpment. *Return to the road by the same path.*

Please do not take a short cut across the boundary wall and wire fence alongside the wood.

Good sections of the limestone can be seen on either side of the road that turns off the A66 towards Penruddock, about 300 m east of the lay-by (432274). These sections were excavated when the new road was cut and they contain abundant fossils, mainly corals and brachiopods. There are two main bands of limestone separated by dark grey shales, and several of the joint surfaces that have been exposed are covered in **tufa**. Between here and Penrith there are some well exposed sections of the Carboniferous Limestone Series along the A66. An idea of the cyclical nature of depositions can be gained by stopping briefly at the old quarry just east of Stainton (496280), and at the roadside section at Redhills (503287). This latter exposure consists of shales, sandstones, siltstones and limestones with a variety of sedimentary structures such as cross-bedding and ripple marks, all of which provide valuable information concerning the environment of deposition.

Location 4: Little Mell Fell (429239). **Basic intrusion into conglomerate**

Little Mell Fell, like its neighbour to the west, Great Mell Fell, is entirely composed of conglomerate. The best exposures on Little Mell Fell are on the southeastern side, for instance just east of the Hause (425235) where the clasts are strongly coated with **haematite**. A worthwhile stop can be made at a small disused quarry just north of Folly Cottage on the track that runs along the base of the eastern side of the fell. Vehicles can be parked just outside the caravan site at Cove.

The quarry consists of a steep-sided intrusion of **olivine basalt** within conglomerate that is very similar to the Pooley Bridge exposures, being made up almost entirely of large clasts of greywacke. Bands of **vesicles** are clearly visible near the steep margins of the basalt. Other intrusions can be found in this locality, and together they seem to indicate a period of basic vulcanicity that perhaps coincided with the eruption of the basaltic Cockermouth Lavas on the northern edge of the Lake District. This period of eruption is generally taken as being of Lower Carboniferous age; but how extensive eruption may have been in the Mell Fell area may never be known, any surface lavas presumably having been eroded away with the rest of the conglomerate.

Location 5: Priests Crag (425233). **Borrowdale Volcanic rocks and panoramic view**

The summit of Priests Crag affords an excellent view of the relationship of the Mell Fell Conglomerate and the surrounding formations. The easiest access is from the Hause, the col between Little Mell Fell and Priests Crag, via the path that leads up the grassy hillside. The road up to the Hause from the previous locality is very steep but there is adequate parking space for a few vehicles at the top.

At the top of Priests Crag are weathered **andesitic tuffs** and lavas, dipping quite steeply to the north-east, and this locality gives the opportunity of confirming that the craggy ground is composed of volcanic rocks. From here one can clearly see the outcrop pattern of the four main formations in this northern Ullswater area:

(a) To the east the low lying areas around the lake are Skiddaw Slates.
(b) To the south and west the BVG forms the rocks of Gowbarrow Fell.
(c) To the north the rounded hills of the conglomerate can be clearly seen, stretching from Great Mell Fell eastwards to Dunmallard at the northern end of Ullswater.
(d) Beyond the northern end of Ullswater, the grey scar of Carboniferous Limestone can be seen forming a feather-edge contact with underlying BVG rocks on the fellside above Roehead (479236).

It is perhaps worth noting a few structural points from this viewpoint (Fig. 10):

(a) The base of the conglomerate can be traced as coinciding with the distinct break in slope that runs from the Hause through Bennett Head to the lakeside just south of Pooley Bridge.
(b) The rugged crags of the BVG are repeated on the other side of Ullswater (Barton Fell, 468214) as the southern limb of the Ullswater anticline. The core of this major fold is composed of softer Skiddaw Slates and has therefore proved to be a comparatively easy line of erosion.
(c) Extensive faulting has complicated the relationship between the Skiddaw and Borrowdale Volcanic Groups, the most notable examples being the Howtown Fault (see Fig. 10) which has led to the lateral offset or 'kink' in the lake itself, and the Ullswater Thrust which separates the BVG from the Skiddaw Slates below.

ADDITIONAL LOCALITIES THAT ILLUSTRATE THE NATURE OF THE SURROUNDING ROCK TYPES

Three localities that enable a more complete picture of this area to be established may be visited.

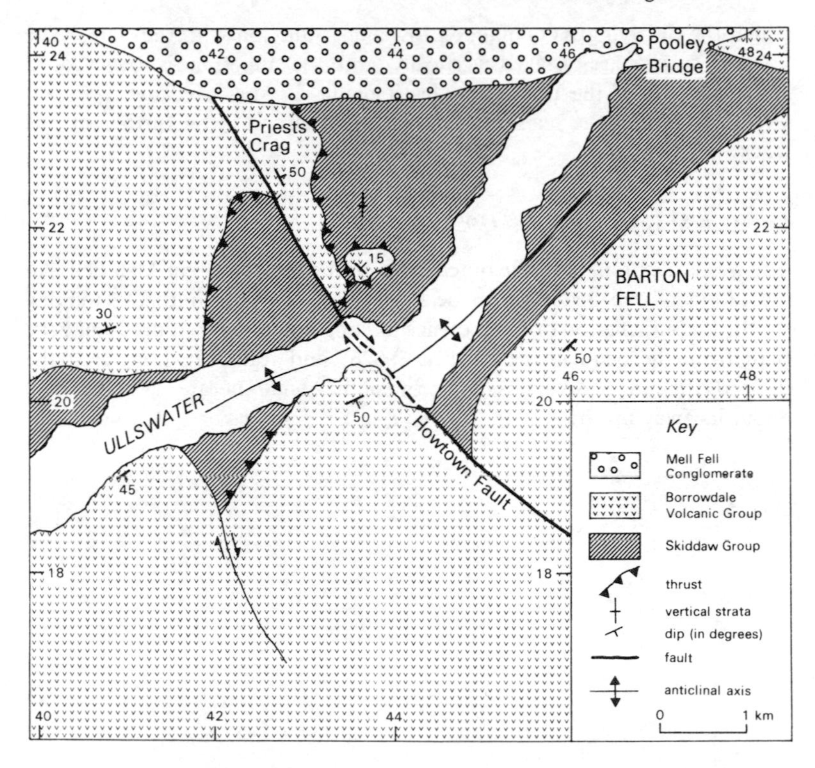

Figure 10 Simplified map of geological formations and structures around north-eastern Ullswater.

Location 6: Stream section in Skiddaw Slates (434223)

About halfway between Watermillock Church and the Knotts caravan site there is a bridge over Pencill Mill Beck. A nearby gate gives good access to the downstream side of the bridge, underneath which almost vertical, rather poorly cleaved dark slaty rocks strike across the stream bed. This is one of the few good exposures of the Skiddaw Slates in this area.

Location 7: Birk Crag (432217)

This locality represents a junction between crumbly, stained Skiddaw Slates below, and cleaved andesitic tuffs and lavas above. Just below the entrance to the Knotts caravan site is a track that leads to an old quarry. Exposures of **'pencil' slates** can be seen first, followed by an interesting section in which the overlying BVG makes a low angle junction with the slates. The volcanic rocks are cleaved vertically and the junction is believed to represent part of the Ullswater Thrust plane. Fresher volcanic rocks are well displayed in the main part of the old quarry a little further up the track.

Location 8: Great Mell Fell (395255)

The western extremity of the outcrop lies at the base of Great Mell Fell. An indication of the bedding can be seen clearly from Troutbeck, looking south. If confirmation of the presence of conglomerate is required, take the path from the bend in the road at 390265 and follow it along past the disused rifle range. An exposed scar of cobble conglomerate can be reached about halfway up the fell.

4 The Armboth Dyke, Thirlmere

(½ day)

Purpose: Examination of the outcrop of the Armboth Dyke, and its displacement due to tear (wrench) faults. If used as a mapping exercise, a whole day would be required.

There is 5 km of walking over rough, steep ground giving on to boggy, undulating moorland, with a gain of some 300 m of altitude, so that adequate fellwalking gear and strong footwear should be worn.

OS maps: 1:25 000 Outdoor Leisure Maps, The English Lakes North-West Sheet. (This shows all the detail required.)
1:50 000 Sheet 90
1:63 360 Lake District Tourist Map

The most convenient access is by car or minibus along the narrow road that follows the western side of Thirlmere. Vehicles should be parked at the small car park at the foot of Fisher Gill (305172). The nearest public conveniences and cafés are in Keswick or Grasmere.

GEOLOGICAL SETTING

The Armboth Dyke is a quartz porphyry rock of a very characteristic and attractive appearance intrusive into the coarse tuffs and agglomerates of the Borrowdale Volcanic Group to the west of Thirlmere. Large phenocrysts of transparent quartz and pink orthoclase appear in a matrix of orthoclase and **oligoclase** which is variously pink or grey. The quartz phenocrysts, generally **euhedral**, typically measure 2–3 mm across. Some of the orthoclase phenocrysts reach a considerable size, often in excess of 1 cm, and **twinning** is not uncommon. A soft material appearing as sparsely scattered small dark patches in the rock (greenish on powdering) is **chlorite**.

The dyke is found to be vertical throughout its entire observable course of approximately 2 km and has a width of 6–10 m. It is **wrench faulted** about seven times along this course, which has a NW–SE trend and extends from an area south of the summit of Fisher Crag (305161) at about

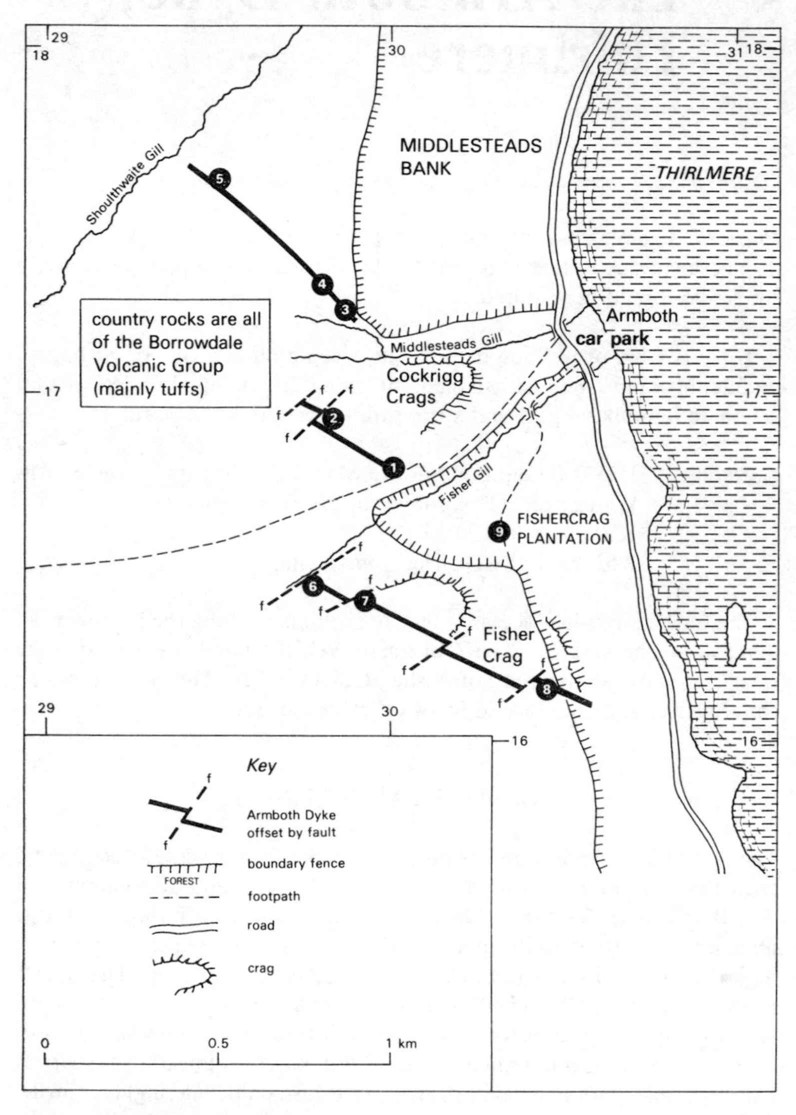

Figure 11 The Armboth Dyke.

385 m OD in a northwesterly direction (Fig. 11). After traversing Fisher and Middlesteads Gills it apparently terminates on the moorland slopes of High Seat towards the head of Shoulthwaite Gill (294176). It does not form impressive topographical features, generally appearing as rounded, ice-scoured knolls. However, its petrology is in such contrast with the surrounding tuffs, and its weathered appearance is so distinctive, that readers armed with a suitably scaled topographic base map (say 1:10 000) may like to try to map the dyke by following its outcrop across the moorland and establishing the points where it is offset by faulting.

Ice from the Thirlmere region has scattered boulders of the dyke over a large area. The present positions of **erratic** boulders of the Armboth Dyke serve to indicate the general direction of former ice movement. Boulders are to be found west and south of the present outcrop scattered over the moorland of Armboth Fell and Wythburn Fell. The fact that perched blocks have been seen on the top of crags 400 m south-west of Harrop Tarn (308133) at a height of some 550 m also suggests a southward movement of glacier ice. Clifton Ward (1876) explains that although the lower ice occupying the Thirlmere valley moved northward, some of the great mass of ice coming off the Helvellyn range shot over the ridge on the western side and escaped partly across the western watershed south of Armboth Fell. As one might expect, the greatest concentration of erratics is found north and NNW of the dyke, in the Shoulthwaite and Naddle Valleys.

EXCURSION DETAILS

Start at the car park at the foot of Fisher Gill, Thirlmere (305172). Proceed on foot along the road in a northerly direction for some 200 m to a gate on the left-hand side of the road by a stream and small stone bridge. This is Middlesteads Gill. Reference should now be made to Figure 11, on which locations of stops have been marked. From the stone bridge at the foot of Middlesteads Gill proceed steeply uphill along the footpath towards Watendlath. Fisher Gill now appears on the left in the trees, and the track starts to zigzag as it reaches a point level with Cockrigg Crags. At around 340 m a cairn is reached at the side of the track consisting of rocks from the dyke **(1)**. Note the characteristic weathering of these dyke rocks in which the quartz phenocrysts have disappeared leaving the rocks full of holes. The loose material and lack of clear contact with country rock of the Borrowdale Volcanic Group give the dyke at this locality an exaggerated width of 16–20 m. Note how the large dyke rocks stand out in relief above the surrounding country here, and also note the small rivulets and damp areas at the edges of the exposure.

Ascend northwestwards along the line of the dyke to a wall and sheep pen. Beyond the wall follow the exposures for approximately 250 m to where a wrench fault shifts the dyke a few metres to the right (**2**). After another 50 m the dyke then disappears due to faulting. Markers for this spot are two circular rings of stones on the moorland a little to the right and below the exposure.

Walk 400 m NNE across Middlesteads Gill to follow the top line of the forest fence where the dyke will again be encountered issuing from the plantation (**3**). In this vicinity are outcrops of green **jointed** and cleaved **tuffs** of the Borrowdale Volcanic Group, and in the dyke itself good doubly terminated quartz phenocrysts up to 5 mm across are to be found.

A hundred metres northwestwards (**4**) marginal dyke rock bearing less quartz and containing more **feldspar** phenocrysts with some **pyrrhotite** and garnet may be seen. Note close-set fractures and rounded, weathered projections. A degree of **thermal** or contact metamorphism giving a reddish hue to **flow-banded** lavas contiguous to the dyke in this area has been observed.

The dyke disappears northwestwards towards High Seat under bracken and peat. The last exposure in this direction (**5**) is at about 460 m at a point before the headwaters of Shoulthwaite Gill are reached (295176). From the Gill go SSE to the wall and sheep pen at **2**, and then take a line SSW (along bearing 200°) to where a footpath crosses Fisher Gill. The dyke reappears as an outcrop, a small knoll (**6**) some 50 m south of Fisher Gill (297164). Here the dyke has been shifted laterally some 400 m to the west by wrench faulting from the place where it was first encountered at **1**. Fisher Gill apparently follows the fault. From here flattish boggy ground has to be traversed for approximately 250 m southeastwards to a pronounced scarp leading to higher ground between 450 and 500 m OD.

The dyke can easily be followed by scrambling up the face of the crag (**7**) (299164), which presents no difficulties. At the top of the crag can be seen the margins of the dyke grading into tuffs of the Borrowdale Volcanic Group. A short rest here will be repaid by good views of Raven Crag and Thirlmere to the north and north-east. The glaciated valley is divided by Great How immediately beyond Thirlmere, and by High Rigg to the north.

Continue along the dyke southeastwards downhill to where it appears under a boggy tarn-filled hollow. Skirt north of the hollow and then move south along the forest boundary to pick up the dyke in a small east-facing crag near the trees (**8**). This is Fisher Crag, at the foot of which are many fallen boulders which would repay examination and from which good specimens can be obtained without damage to the main exposure. A section of the dyke continues into the forest south of Fisher Crag before it is finally faulted out. However, it is not advisable to try and trace it among the trees which hereabouts have suffered much storm damage.

Head north along the forest fence and through a gate to pick up a track leading down through Fisher Crag Plantation. The track is badly eroded as a result of forestry clearance, and at one point **(9)** (303166) storm water has eroded along a fault exposing slickensided and brecciated rock with some reddening and calcite mineralisation. Continue down the track to the road and car park.

5 The Skiddaw Granite in Sinen Gill

(1 day)

Purpose: To examine Skiddaw Granite exposed in Sinen Gill and the effects of thermal metamorphism resulting from its intrusion into rocks of the Skiddaw Group. Old mineral workings will also be seen.

There is 7 km of walking (14 km if starting from Keswick) over bleak and sometimes rough and boggy moorland up to a height of 500 m. Fellwalking gear and strong footwear should be worn.

OS maps: 1:25 000 Outdoor Leisure Map, The English Lakes, North-West Sheet. This shows all the detail required.
1:50 000 Sheet 89 or 90. These sheets can be used, but do not show all the place-names.

The excursion starts at the Blencathra Centre of the Lake District Special Planning Board (303256), formerly a sanatorium and marked as a hospital on the above maps. Access by road is through Threlkeld village (320253) taking the road signposted 'Blease Road leading to Blencathra', but the road is unsuitable for coaches. Cars and minibuses may be parked at the road end by the entrance to the Centre. The area may easily be reached on foot from Keswick via the Brundholme Wood road which runs north of the River Greta. There are shops, cafés and public toilets in Keswick.

GEOLOGICAL SETTING

The Skiddaw Granite forms three separate outcrops set in an extensive region of thermally **metamorphosed** Skiddaw Slates. The outcrops of granite are to be found

(a) in Mosedale where Grainsgill enters the River Caldew (327326), an area described in Excursion 6;
(b) in the vicinity of Wiley Gill, the upper Caldew and Blackhazel Beck (around 310310); and

(c) in Sinen Gill (302282) near the headwaters of the Glenderaterra Beck (this excursion).

That the granite outcrops are connected underground at no great depth has long been supposed due to the fact that the Skiddaw Slates have been thermally metamorphosed over an area of some 70 km^2. The zone of metamorphism extends east to west from Bowscale Fell to Skiddaw and north to south from Coombe Height to Lonscale Fell (Fig. 13). Recent work by the Institute of Geological Sciences has suggested that the Skiddaw Granite is of a **stock**-like form, steep sided and flat topped. This view is supported by M. H. P. Bott, whose work on gravity anomalies envisages a major batholith underlying the whole of the northern Lake District and extending eastwards under the northern Pennines. Radiometric age determinations on the granite give a mean age of 399 Ma, indicating late Silurian or early Devonian emplacement.

The granite is composed of white orthoclase feldspar and **plagioclase** feldspar (oligoclase), biotite mica and quartz. The northern outcrop has been extensively **greisenised** with progressive replacement of the feldspars by muscovite mica of a variety known as gilbertite. This process has been due to **metasomatism** as shown by the development of minerals containing additional fluorine, water and silica not seen in the 'normal' granite, and by the passage of mineral veins from the greisen into the surrounding rocks. The Sinen Gill outcrop contains a rather higher ratio of plagioclase to orthoclase than the 'normal' granite of the central outcrop and so it could be classified as a biotite-granodiorite.

The bulk of the granite is intruded into the Skiddaw Group, a succession of mudstones and greywackes which had previously undergone mild regional metamorphism. The metamorphic aureole extends up to a kilometre from the granite, and previous workers have made much play about the recognition and significance of the various zones of alteration. S. E. Hollingworth in the Sheet 23 (Cockermouth) Memoir of the Geological Survey gives perhaps the best summary.

> In the field the early stages of metamorphism of the Skiddaw Slates are indicated by the development of spotting and **porphyroblasts** of **chiastolite**. With increasing metamorphism there is a progressive hardening of the rock to a massive hornfels, the spots become clearly defined crystals of **cordierite**, the chiastolite changes to clear, translucent **andalusite** and biotite becomes a visible constituent of the base.
>
> The general inward change is thus one of increasing recrystallisation of the hornfels and of increasing definition of characters that appear first in the low-grade type of metamorphism; *there is little that can be described*

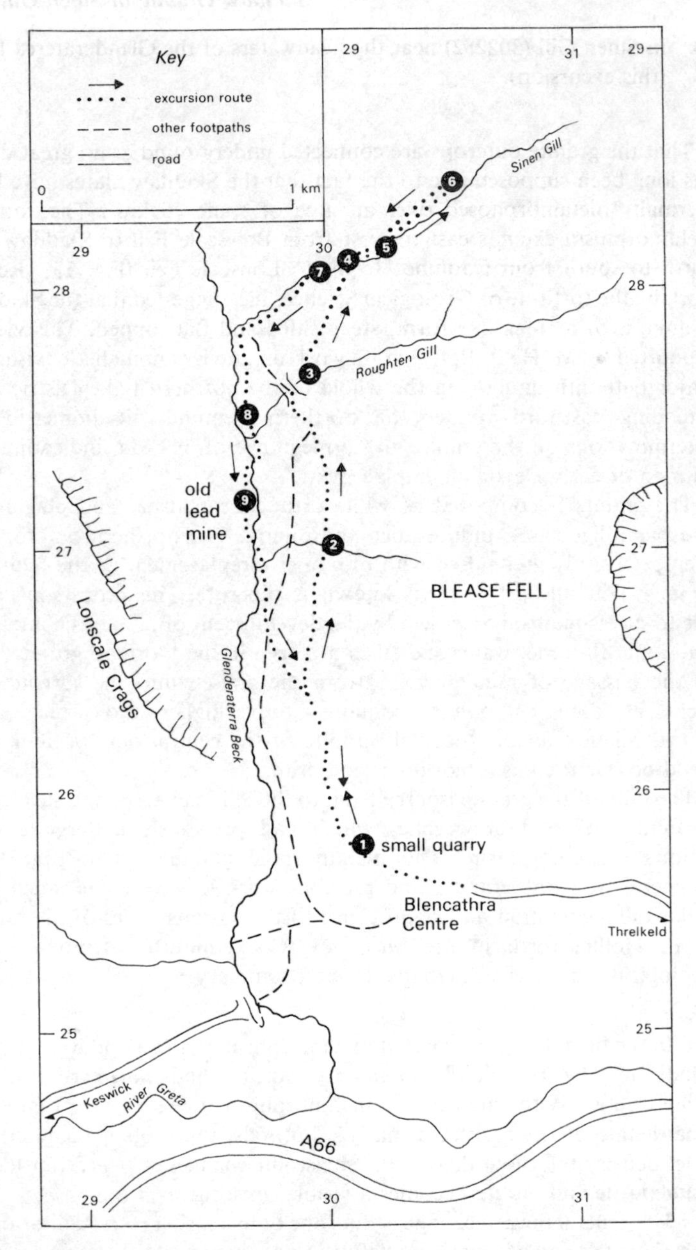

Figure 12 Sketch map of the excursion route to the Skiddaw Granite in Sinen Gill.

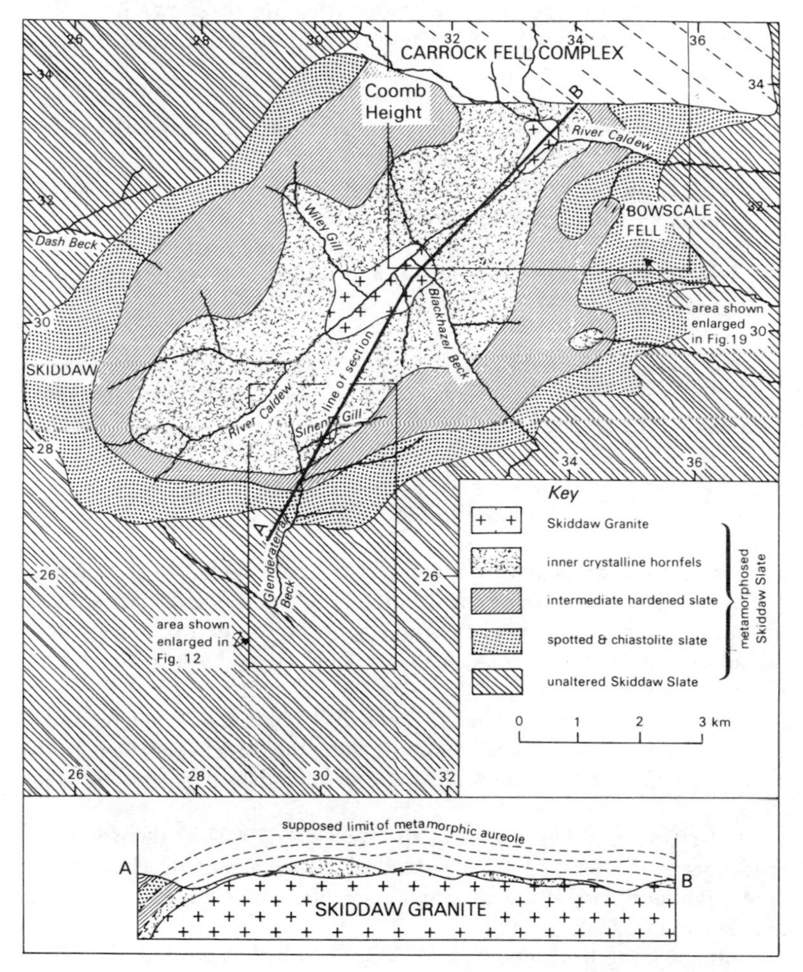

Figure 13 Excursions 5 and 6: geological sketch map and section of the Skiddaw Granite area. (Adapted from *British regional geology, northern England*, by permission of the Director of the Institute of Geological Sciences.)

as zonal appearance of metamorphic minerals (present writer's use of italic and bold print).

Having made this point the Memoir attempts to show, from field observations, three zones of metamorphism of approximately equal thickness:

(a) an outer zone with development of white chiastolite and dark spotting with limited hardening of the rock;
(b) an intermediate zone of considerably hardened hornfelsed slate or mudstone with dull black spots of cordierite; and
(c) an inner zone of tough, well crystallised hornfels with black spots of cordierite having a pitchy lustre, and transparent porphyroblasts of andalusite.

Mineralisation within the area is extensive, with lead and copper-bearing veins following faults and lines of weakness in the country rocks surrounding the granite.

EXCURSION DETAILS

Start from the front gate of the Blencathra Centre (303256) and proceed WNW along the path behind the Centre boundary. Look at the hard blue 'normal' Skiddaw Slate in the cutting opposite the end of the Centre's boundary wall (**1**) (301257). A small overhang at the back of this cutting reveals a slickensided fault plane.

Walk around the lower slopes of Blease Fell keeping to the higher track. Look at the boulders and small outcrops along the sides of the track for the beginnings of thermal metamorphism (ill defined dark grey spotting in the slate).

At the first gill, 1½ km from the start (**2**) (299270), there is much reddening and alteration of the slate with spectacular development of white, randomly orientated chiastolite crystals. It is likely that this gill is aligned along a fault.

Continue on to Roughten Gill (**3**) (298276), which is crossed by a bridge of flat stones. Notice that these consist of hardened hornfelsed slate shot through with blotches of dark cordierite, indicative of the intermediate metamorphic zone. There are also one or two boulders of white Skiddaw Granite in the bridge. Please do not damage any of the bridge stones by hammering. The hornfelsed slate slabs emit ringing tones when *tapped* with a hammer, and selected pieces of this rock have been used for the 'musical stones', a kind of stone xylophone to be seen in Keswick Museum.

Figure 14 Location **4** (301282). Waterfall over Skiddaw Granite in Sinen Gill. (Photograph: T. Shipp.)

From Roughten Gill bear north-east across boggy rising moorland for 500 m until Sinen Gill is reached. Ascend the gill to the prominent waterfall (**4**) (301282) where strongly jointed granite is exposed (Fig. 14). Follow the granite upstream until the granite/hornfels contact (**5**) is seen showing a low angle of dip (Fig. 15). Proceed a further 300 m upstream to some large boulders on the left below a ruined sheepfold (**6**) (305285). Look at the intricate folding with axial plane cleavage in the hornfels, but do not hammer it (Fig. 16).

Retrace your way downstream past the waterfall over the granite at **4** and investigate the deeply weathered, **kaolinised** granite outcrop on the left bank (**7**). Decomposition of the feldspars in the granite may have been accelerated by peaty acids from the moorland, as the joints in the granite here are filled with black, structureless peaty material.

Figure 15 Location **5** (303282). Hornfelsed Skiddaw Slate above Skiddaw Granite. The hammer lies along the contact. (Photograph: T. Shipp.)

Walk down to the confluence of Sinen Gill with Glenderaterra Beck and thence southwards towards Roughten Gill. A mineral vein with abundant iron-stained quartz (gossan) crosses the beck and has been worked at a small collapsed adit. Small specimens of **malachite** and chalcopyrite can be found below another prominent vein at the confluence of Roughten Gill and Glenderaterra Beck (**8**) (296275).

Carry on downstream to the ruins of the Glenderaterra Mine (Fig. 17) on the opposite bank (**9**) (296272). One of the shafts is flooded to surface (Fig. 18), so take care! Abundant specimens of quartz, galena, **pyromorphite** and barite may be obtained.

Re-cross the beck at a point where there used to be a bridge and walk rightwards uphill to the highest track and so back to the starting point.

Figure 16 Location **6** (305285). Small-scale folding and cleavage in hornfelsed Skiddaw Slate. Cleavage almost parallel to hammer handle. (Photograph: T. Shipp.)

Figure 17 Location **9** (296272). The ruins of the Glenderaterra Lead Mine. (Photograph: T. Shipp.)

Figure 18 Location **9** (296272). Flooded shaft at the Glenderaterra Lead Mine. (Photograph: T. Shipp.)

6 The Carrock Fell region

(Two 1-day excursions)

Purpose: The two excursions together allow examination of rocks of the Skiddaw Group, two outcrops of the Skiddaw Granite (complementing Excursion 5), members of the Carrock Fell intrusive complex (gabbro, granophyre and 'diabase'), the Harestones Felsite and Drygill Shales. There are opportunities of collecting a variety of mineral specimens from old mine tips and mineral veins. Corrie formation is exemplified by Bowscale Tarn.

On each excursion there is some 7–8 km of walking on the open fellside, much of it steep, rough, and in places boggy. Fellwalking gear and stout footwear should be worn.

OS maps: 1:25 000 Sheet NY33. This shows all the detail required.
1:50 000 Sheet 90
1:63 360 Lake District Tourist Map
These maps can be used in conjunction with Figure 19.

IGS map: 1:50 000 Sheet 23 (Cockermouth). Drift *or* solid edition.

The first excursion starts at Mosedale (357323), a hamlet with one small shop, which is most conveniently approached from the A66 to the south by the road signposted 'Mungrisdale'. Keswick, Caldbeck and Penrith are the nearest centres with all facilities. Mosedale can be reached by coach, though the road is narrow and has some difficult bends. The second excursion starts at Grainsgill Beck (327327), which can be reached by car or minibus, but coaches should be left at Mosedale because of the problems of turning and parking.

The Carrock Mine is at present (1981) in active operation, mining tungsten ore, and in consequence a small area around it is not accessible to the public. Several old stopes above the Carrock Mine, alongside Brandy Gill, have been re-opened for exploration and mine ventilation purposes. **On no account should these be entered**. Otherwise practically all of the ground can be freely walked over. In the interests of conservation all geologists are earnestly requested to be satisfied with modest specimens and to avoid hammering and spoliation of interesting exposures. In this connection it must be said that even now serious damage has been done at one or two sites.

Two excursion routes are described, Route A taking in the igneous rocks of Carrock Fell, and Route B devoted largely to the mineralisation around Grainsgill. Each route needs a day's fieldwork if it is to be fully covered.

GEOLOGICAL SETTING

The area under consideration lies within a region of lower Ordovician rocks. Sedimentary mudstones and greywackes of the Skiddaw Group occur to the south, while to the north lie lavas and tuffs of the Eycott Volcanic Group known to be older than the Borrowdale Volcanic Group of the central Lake District. Various intrusive masses lie between the sedimentary rocks of the Skiddaw Group and the volcanics of the Eycott Group. First came the **gabbro** complex of Carrock Fell formed about 468 Ma ago, and somewhat later the basic mass now seen as **diabase** was intruded. Later still, at approximately 421 Ma ago, came the more acid Carrock Granophyre and the Harestones **Felsite** (see Fig. 19). The most

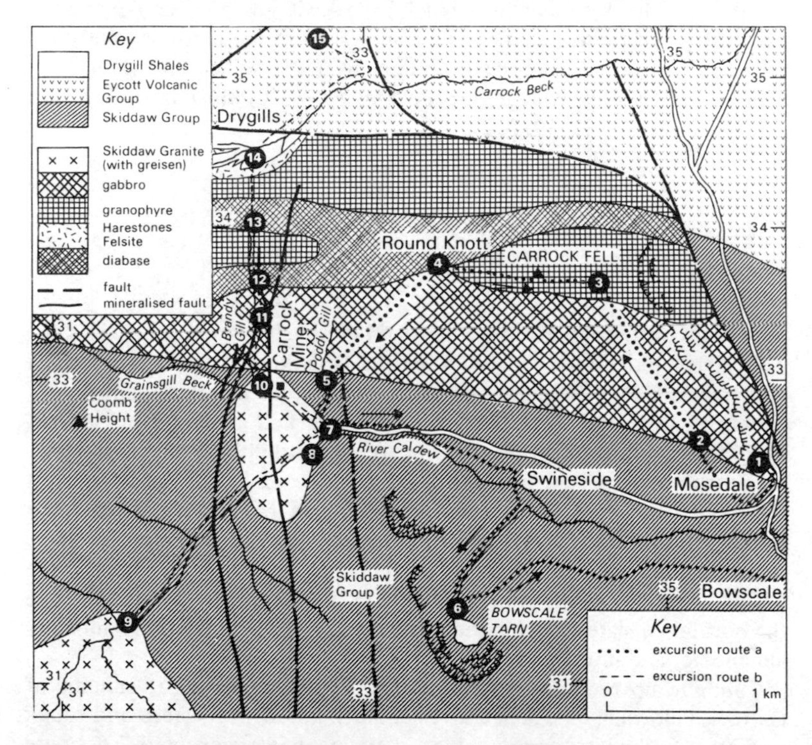

Figure 19 Geological sketch map of the Carrock Fell region.

recent intrusion was that of the Skiddaw Granite, dated at 395 Ma, involving major metasomatic and thermal metamorphic effects in the surrounding rocks and considerable mineralisation along north–south fissures. At a much later date, during **Hercynian** times or later, another suite of minerals was developed to the north, and subsequently there was probably some interaction between the contents of the various veins and their secondary products. As a result the region has a remarkable number of mineral species, and perhaps rather more than half of the known chemical elements are to be detected in one way or another in this quite small area.

This is classic ground for both petrology and mineralogy. It was probably here that Jonathan Otley introduced Adam Sedgwick to the Lake District as early as 1823 when they proceeded from Mosedale up the Caldew Valley at least as far as Wiley Gill. The region has been the subject of many original papers and several excursion guides written from various angles. This present one is aimed at geologists of average knowledge but not having in general access to advanced petrological and mineralogical techniques. The two excursions which follow will give an overall picture of the geology of the area. Further information on the petrology and field relations may be obtained from a paper by I. E. Skillen on the 'Igneous complex of Carrock Fell' in *Proceedings of the Cumberland Geological Society*, **3** (2), 70–85 (1971–2).

EXCURSION DETAILS

Route A

Start from the telephone box at the road junction in Mosedale (357323) and walk 200 m north along the road to a small disused quarry on the left where gabbro is seen **(1)**. The gabbro here is a coarse-grained dark crystalline rock composed largely of white plagioclase feldspar and greenish-black altered **hornblende**, with perceptible amounts of brown **biotite** along the joints.

Return to Mosedale and take the road leading westwards towards Swineside. Just past the last farm building in Mosedale take the path leading up the southeastern spur of Carrock Fell. The path crosses steeply dipping, intensely folded hornfelsed Skiddaw Slates and meets the unfaulted junction with the gabbro **(2)** (352326). Close to the contact zone the hornfelsed slate contains some small garnets which are predominantly almandine, and much biotite.

Continue up the somewhat ill defined track towards the summit of Carrock Fell which is reached after about 2 km of fairly steep and rough walking. On the way various types of the gabbro are exposed. These are coarse, extremely hard speckled rocks predominantly greenish-grey in colour. The so-called quartz gabbro is lightest in colour (leucogabbro) but there are also bands of gabbro richer in ferromagnesian minerals and in the iron–titanium mineral **ilmenite**; these are darker in colour (melagabbro). Banding of the various varieties of gabbros is very marked, and in places xenoliths of altered Eycott Lavas can be found.

On the summit ridge is granophyre, whilst along the margin between the quartz gabbro and granophyre are bands of so-called hybrid rocks where the two types grade into one another. The granophyre is a medium-grained greyish rock which weathers to a characteristic pink tint.

The lower summit of Carrock Fell (**3**) (345336) is the best viewpoint to the east, with the Pennine escarpment and Cross Fell in the far distance and the Carboniferous Limestone scars of Greystoke Park in the middle distance. Below is a former lake basin now occupied by the River Caldew flowing through peaty tracts. At the summit of the fell, which is entirely in the granophyre, there are ice-borne boulders of gabbro lifted at least 60 m

Figure 20 Excursion 6: Bowscale Tarn (location **6**) seen from Carrock Fell. (Photograph: J. D. Hinde.)

above the source rock. In the vicinity the remains of a pre-Roman Iron Age fort may be traced. To the south Bowscale Tarn (Fig. 20) may be seen on the northern slopes of Bowscale Fell (336313) with Saddleback (Blencathra) in the background. This is a good example of a combe-contained tarn in a north-facing position. The only other true combe-tarn in the Skiddaw Group is Scales Tarn, under Saddleback and less than 3 km to the south.

From Carrock summit pass westwards through the 'Main Gate' of the fort and continue in a westerly direction towards Round Knott (**4**) (335337). Here the granophyre is in contact with so-called diabase, altered dolerite, a medium-grained basic rock petrologically distinct from the gabbros but probably of the same general age, whereas the granophyre was intruded later, although before the Skiddaw Granite.

Walk downhill from Round Knott in a southwesterly direction making for Poddy Gill (**5**) (327331). Here, where two little branches join together to form the gill, a quartz vein cuts across just above the junction and from it specimens of schorl, a black variety of tourmaline, have been obtained. This exposure has, however, been excessively ravaged in recent years. A little lower down the gill, about 300 m from the mine road junction, will be seen the spoil heaps of an old lead working on an east–west vein. Lumps of the quartz veinstone may be seen with associated galena.

Having reached the road, turn east and downhill beside the River Caldew. We are now in the metamorphic zone of rocks of the Skiddaw Group, and some very fine examples of pre-folded hornfelses may be seen both *in situ* and in detached masses in the bed of the river. Here also may be seen in some variety in the shingle beds fragments of many of the rocks of the upper Caldew valley, including Skiddaw Granite, the mica-rich greisen, and variously metamorphosed rocks of the Skiddaw Group such as spotted slate, cordierite and andalusite schists.

The excursion may now be terminated by walking down the valley road to Mosedale (3 km). The more energetic, however, may like to cross the river at Swineside and climb rather steeply up to get a close view of Bowscale Tarn (**6**) (336314) where the rocks are still well within the metamorphic aureole of the Skiddaw Granite. An easy return may be made from this point down the track to Bowscale with fine views of the U-shaped valley of the Caldew. From small exposures of unaltered black Skiddaw Slate above Bowscale the diligent searcher for graptolites may be rewarded. The round is completed by walking across to Mosedale. The 'new' bridge at Mosedale is a pleasing example of the use of hornfels as building material.

Route B

The starting point for this excursion is Grainsgill bridge (**7**) (327327) which can be reached by car or minibus (*not* by coach). Almost immediately below the bridge is the contact between the partially greisenised, highly micaceous and quartzose form of the Skiddaw Granite and highly altered Skiddaw Group hornfels. Just above the bridge and to the north of the stream it is possible to see a narrow zone where the hornfels itself has also been converted into greisen.

Now cross the bridge and proceed down to the River Caldew where the granite may again be seen, here in its normal form. It is a light coloured rock showing white feldspars and quartz with some black mica. As it is exposed in the river bed the outcrops are not very accessible but it is possible to see at the western end of the exposure the contact with altered rocks of the Skiddaw Group **(8)** (322322). The contact is almost horizontal, showing the relatively recent unroofing of the intrusion. If desired, more time could be spent in proceeding up the valley as far as the main central outcrop of the granite **(9)** (315314), with more examples of rocks from the metamorphic aureole including cordierite and andalusite schists between the two granite outcrops. A map and section of the three outcrops of Skiddaw Granite is shown in Figure 13.

Return to Grainsgill Bridge and proceed up past the Carrock Mine to **10** at the foot of Brandy Gill (323329). One should first notice the fine display of north–south veins coming down from Coomb Height, crossing Grainsgill and penetrating the southern slopes of Carrock Fell. They are predominantly quartz veins carrying a rather remarkable variety of metallic minerals. Three of these veins have been, or are being worked: from east to west they are the Emmerson, the Harding and the Smith veins. There are also two veins in fault fissures containing ankerite (p. 56), while numerous veins and strings show traces of mineralisation. It is not now very practicable to obtain specimens of metallic minerals from the actual veins, but there are extensive spoil heaps which represent the residue of working at several periods since early this century until the present day. Naturally these heaps have been picked over innumerable times but specimens, although generally small, may still be found.

The chief requirements are patience and some knowledge of what to look for. The following list indicates *metallic-looking* minerals, i.e. those with a metallic lustre, which might be seen. Note that the hardness of the minerals is given on Mohs' scale. For practical purposes a geological hammer head or knife blade is about 5½ on Mohs' scale, a 'copper' coin about 3½ and a finger-nail about 2½. A harder mineral will scratch a softer one unless it is very brittle.

Galena, PbS	Blue-grey, cubic cleavage; hardness 2.5; heavy
Pyrite, FeS_2	Pale brassy-yellow; hardness 6–6.5
Chalcopyrite, $CuFeS_2$	Brassy yellow; hardness 3.5–4
Pyrrhotite, Fe_7S_8	Bronze tarnish; magnetic; hardness 4
Arsenopyrite, FeAsS	Tin-white, often tarnished black; garlic odour when hammered, hardness 5.5–6
Sphalerite, ZnS	Usually the dark ferriferous variety, Fe substituting for some of the Zn; glistening black sub-metallic to resinous lustre; pale streak; hardness 3.5–4

Molybdenite, MoS_2	Steel-blue flakes; soft and flexible; hardness 1–1.5
Wolfram, $(Fe,Mn)WO_4$	Coal-black sub-metallic lustre; tabular; heavy; hardness 5
Bismuth, Bi	Granular; silvery, with characteristic reddish tint; brittle; hardness 2–2.5
Bismuthinite, Bi_2S_3	Steel-grey, acicular crystals; fragile; yellow tarnish; hardness 2
Joseite, Bi_4TeS_2	(Formerly identified as *Tetradymite*). Brilliant silvery white, with perfect cleavage; hardness 1.5–2

The bismuth minerals are generally rare.

Non-metallic minerals are also numerous but often less easy to pick out. The following may be looked for:

Quartz, SiO_2	Generally milky-white and abundant, but good small hexagonal crystals should be looked for; hardness 7
Calcite, $CaCO_3$	Also milky-white and crystalline; hardness 3; it cleaves into rhombs and fizzes with acid
Dolomite, $(Ca,Mg)(CO_3)_2$	Usually pinkish-cream in colour; satiny lustre; hardness 3.5–4
Ankerite, $2CaCO_3.MgCO_3.FeCO_3$	Largely fills the 'ankerite veins', weathering dull brown to produce umber; hardness 3.5
Apatite, $Ca_5F(PO_4)_3$	Usually in pale green hexagonal prisms on quartz; hardness 5
Fluorite, CaF_2	Small amounts in quartz; generally purple colour; hardness 4
Scheelite, $CaWO_4$	Creamy brown, sometimes showing tetrahedral faces; associated with wolfram; gives violet fluorescence with ultraviolet lamp in the dark; hardness 4.5–5
Gilbertite	A white, hydrous mica found as glistening crystals usually along the selvedges of quartz veins in greisen; hardness 2–2.5

From this mineralogically rich area it is now convenient to proceed north up Brandy Gill. At the lower waterfall, about 10 m high **(11)** (323334), is an old level and some exposure of quartz veins on the right bank of the beck.

On the left bank there is an outcrop of one of the ankerite veins which here contains the pale green amorphous mineral scorodite, an iron arsenate mineral no doubt derived from the oxidation of arsenopyrite. Unfortunately much of this outcrop has now been removed.

Continue up Brandy Gill where one may see the continuation of the north–south veins penetrating the various Carrock Fell intrusives, which are here in a very weathered condition. Above the second waterfall **(12)** (323336) there are more old mine workings on east–west veins cutting across the gill. From the spoil heaps of these various secondary minerals have been obtained, including rare tungstates, molybdates, arsenates and phosphates of lead and copper. Discovery and recognition of these requires considerable knowledge and experience, and specimens will probably be minute.

Climb to the head of Brandy Gill **(13)** (323340), where east–west veins are exposed with 'red gossan' and contain the black hydrated oxide of manganese, psilomelane.

Proceed northwards over the stretch of moorland to drop down into Drygill **(14)** (323345). Here will be seen a small faulted exposure of Upper Ordovician Drygill Shales. The rock is blue-grey, weathering white and containing fragmentary trilobite fossils, mainly trinuclids.

Just north of the shales is a strong east–west vein associated with the main Caldbeck Fells mineralisation and famous for the mineral campylite which is an arsenical pyromorphite $Pb_5(PO_4)_3Cl$, occurring as yellowish-brown barrel-shaped crystals. Actually much of the abundant pyromorphite of the Roughton Gill veins contains arsenic, and the amount seems to decrease progressively as one proceeds in a westerly direction.

From Drygill the return journey may be made over the same ground or a further extension north to Driggeth **(15)** (327352), where the main Roughten Gill lode is exposed at the surface, may be made.

7 The Carboniferous Limestone between Caldbeck and Uldale

(½ day)

Purpose: To examine part of the Carboniferous Limestone succession on the northern edge of the Lake District.

This may best be taken as a part-day excursion doing the 'round of Skiddaw' by car or coach from Keswick, passing through Mungrisdale (364304), Hesket Newmarket (340386), Caldbeck (324397), Uldale (249370), Castle Inn, Bassenthwaite (216326) and so back to Keswick. The locations described are freely accessible and are only short distances from the roadside on open moorland.

OS maps: 1:50 000 Sheets 90 and 85
1:63 000 Lake District Tourist Map
IGS map: 1:50 000 Sheet 23 (Cockermouth), solid *or* drift editions
The Faulds Brow Quarry is shown only on sheet 85 and on the IGS map, the latter showing all localities.

Shops and toilets are to be found in Caldbeck village. Parking is possible on roadside verges and lay-bys adjacent to the localities described.

GEOLOGICAL SETTING

The area immediately north of Caldbeck–Uldale forms a tilted, dissected plateau with beds of limestone, sandstone and shale dipping north-west away from the fells. Carrock Fell, with its unique petrology and mineralisation (described in Excursion 6) stands out boldly to the south whilst southwestwards the end-on view of the summit ridge of Skiddaw makes a bold spire.

The Lake District is almost completely encircled by outcrops of Carboniferous Limestone, occurring as tilted fault blocks dipping away from the mountains of the central part. These Lower Carboniferous rocks show a very varied nature from place to place around the perimeter of the

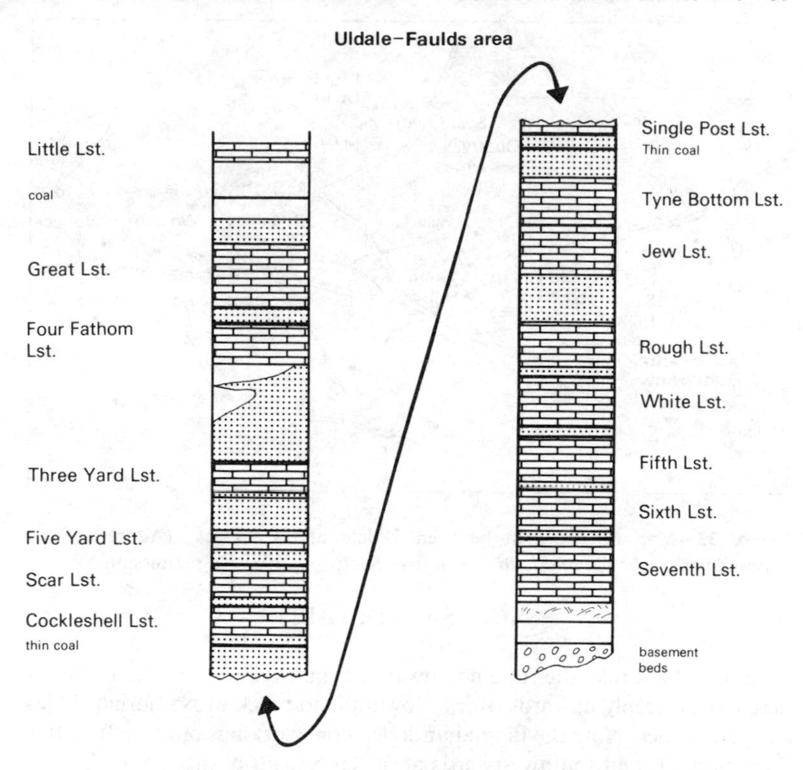

Figure 21 Carboniferous Limestone succession in the area between Uldale and Caldbeck (From *Proceedings of the Cumberland Geological Society* **4,** 47, with permission.)

District. For instance, here in the Caldbeck region they show a lithology transitional between the Lower Carboniferous of West Cumbria and that of the Alston Block in the North Pennines; perhaps this is not entirely surprising in view of the location. The tendency is for the West Cumbrian limestone units to thin out northeastwards, whilst the intervening shales and sandstones assume more significance. As we move towards the north and east the limestone lithology becomes more estuarine. The lithology of the Uldale–Caldbeck limestones is shown in Figure 21.

Three localities have been chosen to illustrate the geology of the area. These are numbered on Figure 22, which also indicates additional localities which can be investigated if time and inclination allow.

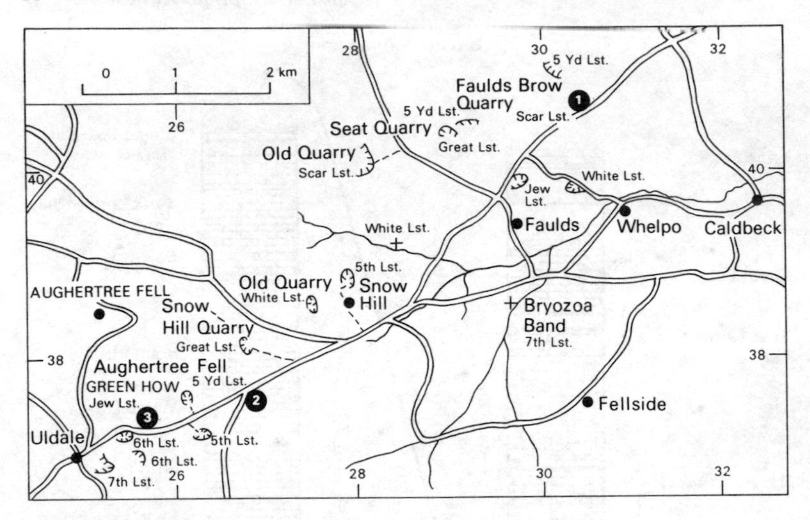

Figure 22 Map of the area between Uldale and Caldbeck. (Adapted from *Proceedings of the Cumberland Geological Society* **4**, 49, with permission.)

EXCURSION DETAILS

From Caldbeck take the road northwards signposted to Carlisle. The road here passes steeply upwards over a downfaulted block of Namurian shales and sandstones. Note the long-abandoned coal workings on the left. After 2 km turn left and southwestwards at the crossroads at the top of the hill, and after a further 2 km Faulds Brow Quarry may be seen on the right **(1)** (304406). This is in the Scar Limestone (Figs 21 and 22). A generalised section through the quarry is shown in Figure 23.

Walk into the lowest part of the quarry, which has been worked along a bedding plane dipping at 12° WSW. The main wall of the quarry consists of pale grey compact limestone about 8 m thick showing widely set bedding planes and strong joints. It is not very fossiliferous.

The upper part of the quarry can be approached via an earth slope at the left-hand (western) end. **On no account should the quarry face or the remains of the rock chute be climbed** as the rock has been loosened by explosives. In the floor of a shallow embayment at the left-hand end of the upper face can be seen compact grey granular limestone with rusty-weathering nodules of pyrite. The limestone throughout the quarry is strongly jointed and it is worth noting the chemical action of groundwater percolating into the joint blocks and staining them to a depth of several centimetres. Also, along the joints, solution and precipitation of calcite has taken place. The walls of this shallow embayment consist of 1 m of grey,

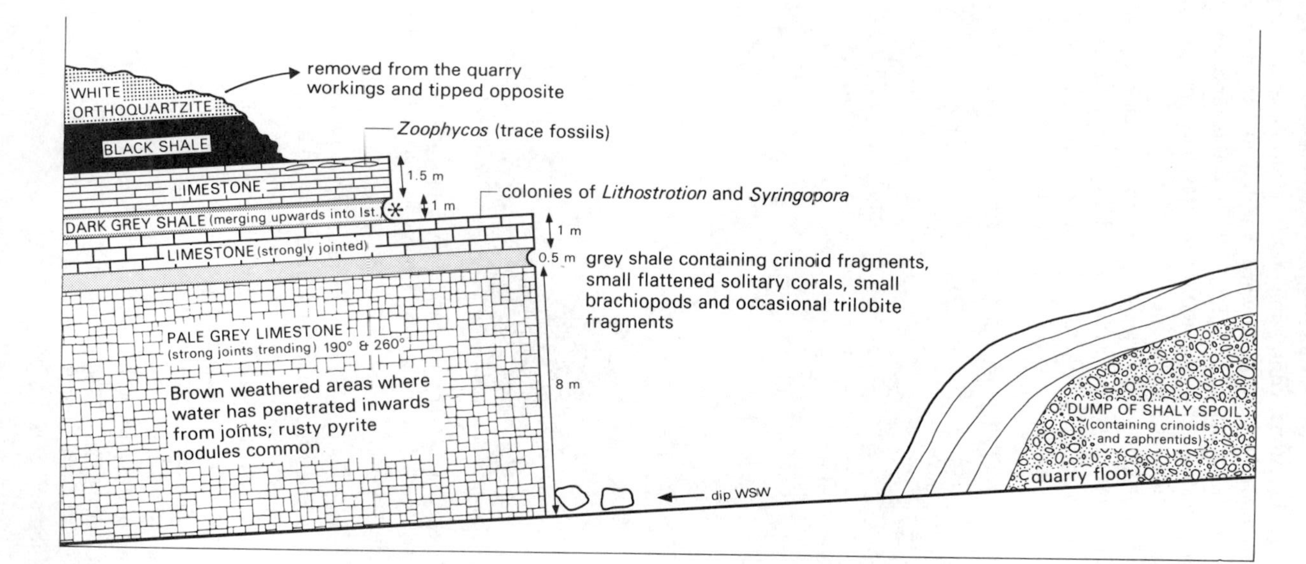

Figure 23 Generalised section through Faulds Brow Quarry (location **1**) near Caldbeck. The burrows at the horizon marked with an asterisk (*) form one of the most interesting features of this quarry as they are clearly aligned below joints in the limestone and are therefore small solution channels: a cave system in miniature.

rather gritty limestone containing some crinoids and solitary corals (notably *Aulophyllum*), whilst towards the top, 0.4 m of dark, crumbly shale is full of fragments of crinoids, solitary and compound corals, small brachiopods, bryozoa and occasional trilobites. The shale merges upwards into a coarse fragmental limestone.

Work several metres rightwards up on to a small area that has been loosened by shotfiring. On weathered bedding planes can be seen curious spiral fan-like markings of the trace fossil *Zoophycos*, thought to be the feeding burrows of an unknown marine organism (Fig. 24). Scramble

Figure 24 Location **1**. The Carboniferous Limestone between Caldbeck and Uldale. Trace fossil *Zoophycos*. Scar Limestone, Faulds Brow Quarry (304406). (Photograph: T. Shipp.)

carefully down (about 2 m) on to the main upper bench of the quarry. The upper 2 m of limestone had clearly been worked along a joint plane. The second set of joints, almost at right angles to the face, can be seen as fine vertical cracks. At the foot of each joint, at the shale band which has obviously held up the passage of groundwater, can be seen small 'burrows' 10–20 cm in diameter which form an underground network, a cave system in miniature (Fig. 25).

The floor of this part of the quarry, which is a gently undulating bedding plane, can be envisaged as once part of the bed of a tropical Carboniferous sea. Here and there can be seen patch reefs of fossil corals, including *Lithostrotion* and *Syringopora*.

Figure 25 Location **1**. The Carboniferous Limestone between Caldbeck and Uldale. A cave system in miniature, Faulds Brow Quarry (304406). (Photograph: T. Shipp.)

The quarry backs on to open moorland, and from here beds of dark grey fossiliferous shale succeeded by pale brown **orthoquartzite** with fossil burrows have been removed and dumped adjacent to the quarry entrance. This dump of spoil furnishes good specimens of the ram's horn coral *Zaphrentis*, crinoid stems and plates, and small brachiopods.

From Faulds Brow continue 5 km southwestwards along the road towards Uldale. Stop at any convenient roadside lay-by in the region of Aughertree Fell (268377) and investigate the moorland south of the road (**2**). A number of depressions up to 10 m deep will be seen. These are swallow-holes in the limestone, here buried under peat and drift cover. It may well be that these depressions are aligned along major joint intersections or faults.

A further 1 km on, where the road crosses Green How (258373) there are some grassed-over spoil tips a few metres uphill to the right (**3**). This long disused quarry is in the Jew and Tynebottom Limestones, and the following sequence may be followed with some difficulty. At the base, about 2 m of pale grey limestone with reddish spotting contains the coral *Lonsdaleia* and large brachiopod *Gigantella*. Above is 6 m of well bedded, rather splintery limestone with *Saccaminopsis*, a small bead-like foraminiferid, and the corals *Aulophyllum*, *Dibunophyllum* and *Nemistium*. High up in the quarry is an excellent marker band 15 cm thick containing the compound coral *Lithostrotion phillipsi*, whilst at the top of the succession is 1 m of dark limestone containing *Caninia*. A few specimens may be obtained from the rather sparse scree lying about, but fossils found *in situ* should be left there for others to see.

There are many more old quarries and limestone and coal diggings in this interesting area, as indicated on the sketch map (Fig. 22), but space does not permit their enumeration here. Readers interested in making further investigations will find some useful notes compiled by Mr E. H. Shackleton in *Proceedings of the Cumberland Geological Society*, **4** (1), 46–50 (1977–8).

8 Borrowdale

(1 day)

Purpose: To view the effects of glaciation, including *roches moutonées*, a marginal meltwater channel, terminal moraines and a possible perched block; and to examine rocks of the Skiddaw Slate Group and the Borrowdale Volcanic Group, including the junction between these groups.

There is 11 km of walking, mainly on good paths, requiring about 7 hours; this may be shortened to 7 km by omitting **6** and **7**. Fellwalking gear, including boots, is recommended.

OS maps: 1:25 000 Outdoor Leisure Map, The English Lakes, North-West Sheet (This shows all footpaths and place names.)
1:50 000 Sheet 89 or Sheet 90
1:63 360 Lake District Tourist Map

The excursion starts from Keswick. All roads referred to are suitable for cars, but coaches should avoid the narrow road west of Derwentwater, using the B5289 instead. The public bus service from Keswick to Borrowdale passes Quayfoot Quarry (**1**). Toilet facilities are available in the Quayfoot car park (253168) and in Grange village (253174). There are cafés and shops in Grange, and a village store and hotel with public bar in Rosthwaite.

GEOLOGICAL SETTING

Borrowdale can be defined as the drainage area of the River Derwent south of Derwentwater. It includes some 60 km^2 of spectacularly beautiful countryside combining high fells with soaring crags, steep woodlands of mainly hardwood trees, and flat meadows with clear, fast-running streams. Derwentwater itself and the fells to the west are in the Skiddaw Slates, but the junction with the overlying Borrowdale Volcanic Group runs down the eastern shore of Derwentwater, crosses the valley just south of the lake, and rises up the steep west wall to the south-west of Grange village. This has resulted in some marked differences in topography between the hard volcanic rocks of the Borrowdale Group and the older but softer metamorphosed shales and grits of the Skiddaw Group. Nowhere is this seen more dramatically than at the entrance to Borrowdale where the wide

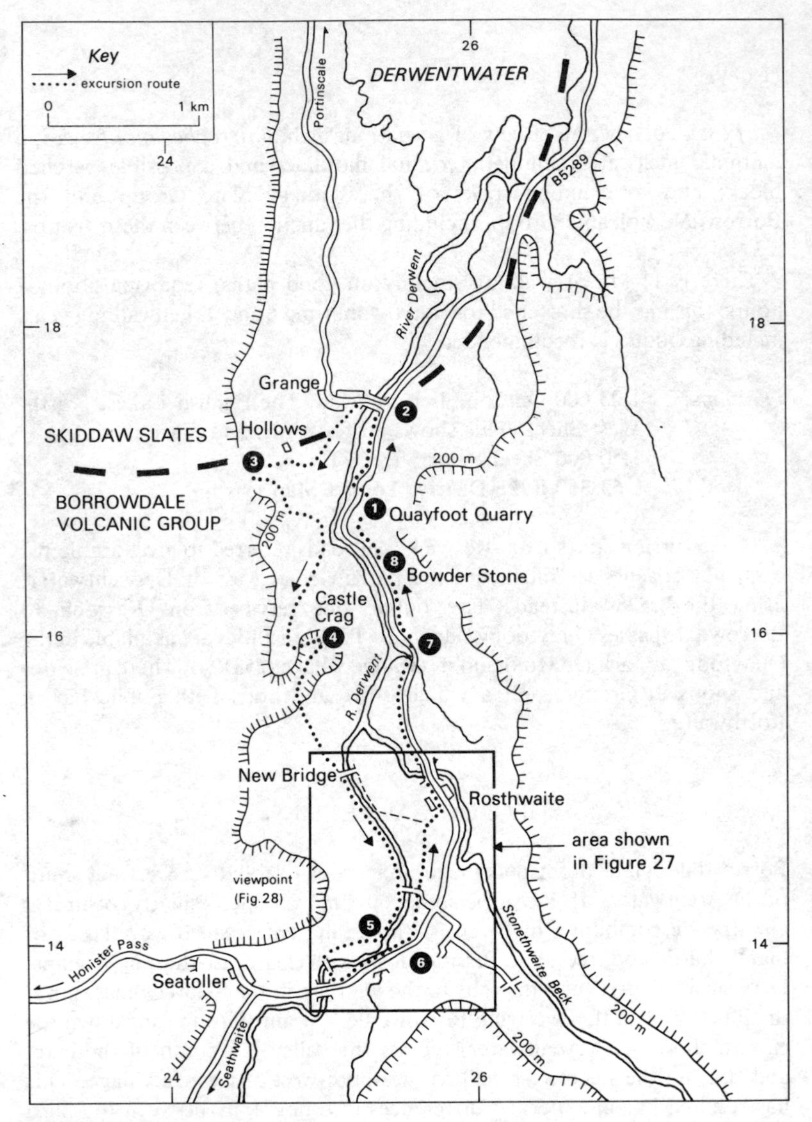

Figure 26 Borrowdale.

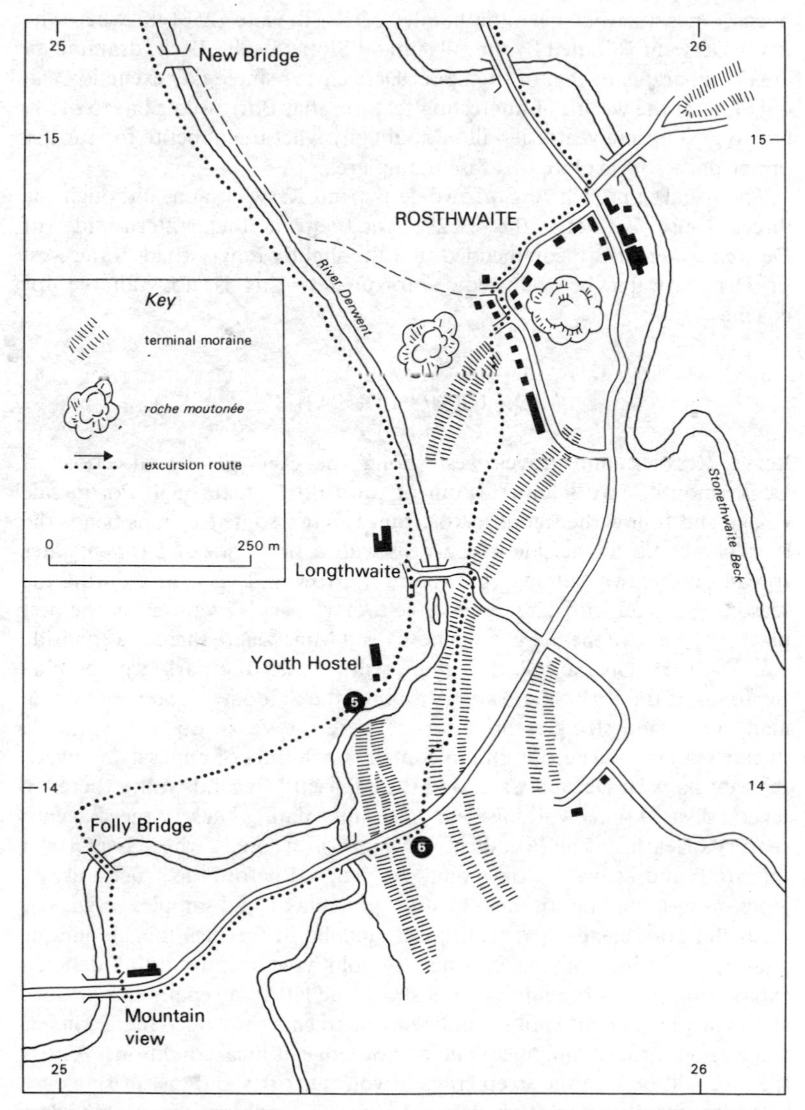

Figure 27 Rosthwaite and surrounding area showing the principal glacial features.

flat valley of Derwentwater is constricted into a narrow, steep-walled gorge immediately south of Grange village (252175). Farther south the valley opens out somewhat near the hamlet of Rosthwaite (258148) where the River Derwent is joined by the substantial Stonethwaite Beck, draining an area comparable to that of the upper Derwent. A single-day excursion can only hint at the wealth of interesting features that Borrowdale has to offer, but to a first-time visitor it will undoubtedly whet the appetite for further opportunities to explore this fascinating area.

The usual approach to Borrowdale is from Keswick and, although the direct route is along the B5289 road down the eastern side of Derwentwater, it is recommended that the slightly longer road to the west of Derwentwater be followed; unfortunately this is not suitable for coaches.

EXCURSION DETAILS

Leave Keswick and travel west along the A66 in the direction of Cockermouth. Two kilometres out of town turn left through Portinscale village and follow the signposts to Grange. After some tortuous bends the road climbs up to the side of Catbells with a fine vista of Derwentwater spread out below. Although the road is narrow and very busy during the season, it should with care be possible to park for a few minutes, the best place being above the old Brandlehow Lead Mine where there is a roadside seat (249195). On the fellside behind will be seen the dark grey or blue siltstones of the Kirk Stile Slate division of the Skiddaw Group containing many veins and stringers of quartz. Where the veins attain a workable thickness they have been found to contain a rich suite of minerals. Between the west bank of Derwentwater and the adjacent Newlands valley there are several disused mine workings and smelt mills dating back to the sixteenth century or earlier. The Brandlehow Mine, for instance, whose overgrown tips are found below the viewpoint, was worked before the general use of gunpowder in mining. In these tips it is possible to find samples of galena, **cerussite**, **zinc blende**, pyrite, fluorite and barite besides the ubiquitous quartz, and it is recorded that traces of gold have been found. A detailed exploration of this old mining area should be left to a separate excursion, and in any case no attempt should be made to enter any open mine adits or pits without proper equipment and a knowledge of local conditions. Across the lake will be seen the steep crags of volcanic tuffs and lavas rising very abruptly from the eastern lake shore along which the Borrowdale Volcanics/Skiddaw Slate junction runs for some distance.

If visibility is good to the north and north-east, the much more rounded outlines of Skiddaw and Blencathra may be seen, formed of rocks of the

Skiddaw Group. Continue down the road into Grange village noting the rocky knolls or **'roches moutonnées'** dotting the silted-up valley floor. Cross the River Derwent by the picturesque double-spanned bridge and turn right (south) along the B5289 road for 0.5 km to **1** at Quayfoot car park (253168) where vehicles will be left for the day. If travelling by bus from Keswick, alight at Grange Bridge (254175).

Walk back along the road to Grange (**2**) and while crossing the bridge note the fine *roche moutonnée* on which the western end of the bridge has been built, the rock having been smoothed and deeply scored by the pressure of ice grinding over it for many thousands of years. Opposite the Grange Hotel turn left (south) along a narrow lane and within a short distance the view of the western side of the valley opens out and the white buildings of Hollows Farm will be seen. Above and to the right of the farm the fell slopes are smooth with few outcrops of rock, indicating the softer Skiddaw Slates, but to the left the basal Borrowdale Volcanics rise in a series of great rocky steps sweeping up to the skyline. At the lane junction turn right and walk up to Hollows Farm (247171) where, after requesting permission, access can be gained to the fell side immediately behind the farm. Climb the steep grass slope alongside a marshy stream and after reaching some outcrops of Skiddaw Slates on the right there is a break in the slope, whilst up and to the left are the crags in which will be found dark grey but whitish weathering volcanic tuffs of the Borrowdale Volcanic Group and the underlying brown Skiddaw Slates (**3**).

A further climb up the line of crags (246170) will reveal good exposures of the tuffs and lavas and there are excellent views both north and south illustrating the differing topography of the two main rock groups. Here and at any other natural rock outcrops hammering should be restricted to an absolute minimum. There is always plenty of loose material close to the outcrops to produce fresh surfaces for examination or collection. The indiscriminate hammering of rock outcrops is usually not only ineffective in yielding good specimens but also destroys the beautiful weathered features which have taken thousands of years to develop and which cannot anyhow be seen in small hand specimens.

Turning downhill, with Hollows Farm behind to the left, bear right above Scarbrow Wood and follow the path down to a metal stile and a clearing alongside the river (250167). The old packhorse trail now leads south away from the river towards the gap between the rocky spire of Castle Crag and the steep fells to the west. This gap is a good example of a marginal glacial meltwater channel and it can be seen to advantage from the top of Castle Crag. The latter is now climbed by turning left at the National Trust sign close to a big cairn and making the steep but easy ascent to the summit (**4**) (249159). The upper part of this pinnacle has been extensively quarried in the past as it consists of a hard, fine-grained

Figure 28 Borrowdale. A view northeastwards from High Doat (247145) showing the constricted gorge of the River Derwent at the 'Jaws of Borrowdale' (left). Grange Fell (opposite) is composed of tuffs and lavas of the Borrowdale Volcanic Group. (Photograph: K. W. Bond.)

volcanic tuff with a well developed cleavage making it very suitable for roofing slates. The views from the top, which is reputed to be the site of a Romano-British fort, are quite spectacular. Below to the east is the River Derwent flowing through the deep constricted gorge known as the 'Jaws of Borrowdale' (Fig. 28). This widens out northwards at Grange village into meadows and marshes which were once the southern part of Derwentwater, the river now entering the lake through a nice 'bird's foot' delta.

Southwards the valley opens out again into a roughly diamond-shaped area around Rosthwaite which was once a lake dammed by moraines in the narrow opening to the 'jaws'. The hamlet of Rosthwaite, a Norse settlement like most others in this valley, was located on the sheltered side of a cluster of *roches moutonnées* slightly above valley level to escape the periodic flooding of the old lake bed. Beyond Rosthwaite the valley forks into two main branches with the Stonethwaite valley to the left and the main Derwent valley leading up to Seathwaite on the right.

Leave Castle Crag by the same route but after descending 70 m over quarry debris go south along the ridge to a stile visible in a clump of Scots pines, and after keeping to the high ground for another 300 m bear south-east and descend to the valley at New Bridge (251151). At this point it is possible to shorten the excursion by crossing the bridge and turning right to walk into Rosthwaite where the bus to Keswick can be caught or the road taken northwards back to the car park at Quayfoot. If time and weather permit, do not cross the bridge but take the footpath on the west side of the river to Longthwaite Youth Hostel (255143), then continue for a further 150 m to a point where the path drops down to the water's edge by a bend in the river opposite a high bank of eroded moraine (**5**). To appreciate the features of this area it is necessary to consider the final stages of the last glaciation, probably some 12 000 years ago, in more detail (Fig. 27). At that time Borrowdale was occupied by ice originating from two vigorous, Alpine-style glaciers flowing down the Stonethwaite and Seathwaite valleys.

As the ice in the area melted away the Seathwaite glacier melted faster so that during one of the many temporary respites in the general melting the Stonethwaite glacier was blocking the Derwent valley around Longthwaite and Rosthwaite while the Seathwaite glacier ice front had retreated back up to Thornythwaite (246134) or beyond. As a consequence meltwater from Seathwaite formed a lake held back by the Stonethwaite glacier, and the upper River Derwent could only escape on the extreme western side of the valley along a marginal channel between the fellside and the glacier. From our viewpoint it is possible to see how the river has gradually cut down through moraine, draining away the lake and leaving behind the flat lake bed on the upstream side. Underfoot, and as high as 3–5 m above the present water level in the river, are found water-abraded potholes in the rocky outcrop of the west bank indicating the earliest level of the overflow channel. The moraine on the opposite bank is a mixture of graded sand beds and completely unsorted glacial till containing distinctive igneous rocks from the Stonethwaite glacier drainage area. When this glacier ice front resumed its retreat it did so in stages, leaving behind two or possibly three concentric terminal moraines crossing the valley, the outermost being the one already seen. Readers wishing to make a more exhaustive study of **5** and **6** are recommended to read an article by P. Wilson on 'The Rosthwaite Moraines' in *Proceedings of the Cumberland Geological Society*, **3** (4), 239–49 (1975–6).

Continuing the excursion, follow the path south-west for 400 m through Johnny's Wood (a National Trust oak forest) to Folly Bridge and join the main road at Mountain View (251137). Turning left and progressing north-east for some 200 m, there is a good view of the outermost moraine running right across the valley and marked by a line of trees. A stream from the east side of the valley flows along the foot of the moraine and after being crossed by the road at Burthwaite Bridge (254139) joins the River Derwent. This has also been diverted across the valley floor to the far west side where it breaches the moraine at **5**. Immediately beyond Burthwaite Bridge the

Figure 29 Borrowdale, location **7**. *Roche Moutonnée* at Red Brow (256159), sectioned during road building. (Photograph: K. W. Bond.)

road surmounts the moraine. 100 m beyond the moraine turn left off the road at the signpost 'Footpath to Rosthwaite' (**6**) (256140). Across one field the path climbs to the top of the middle moraine, which is high enough to give a good view of this and the other moraines swinging round towards Rosthwaite. The path joins the road at a house aptly called 'Moraine', but after 70 m leaves it again at 'Peathow' to cross several fields to the village. Before reaching the houses the outermost moraine is encountered again. On either side are outcrops of *roches moutonnées*, smoothed and striated on the southern, upstream sides but steeper and craggy on the downstream sides due to the plucking action of the ice. Join the main road at Rosthwaite and proceed northwards for 1 km across the flat stretch of old lake bed to **7** at Red Brow (256159) where the road enters the 'Jaws of Borrowdale'. Road-building has truncated a *roche moutonnée* to show a nice section of fresh rock with a bank of glacial drift above containing stones of all shapes and sizes (Fig. 29). Around the corner are some vertically bedded andesite lavas with amygdales of calcite, chlorite or quartz.

Take the signposted path off to the right leading up to the Bowder Stone (**8**) (254164), an enormous boulder calculated to weigh 2000 tonnes and seemingly balanced on a sharp edge. This has either rolled down from the slope above or more likely was transported by ice and gently lowered when the glacier melted back. Continue down the path to Quayfoot car park nestling among trees planted to landscape the old quarry tips. On fresh surfaces the rock is seen as a green, fine-grained **tuff** with high-angle cleavages crossing the original bedding. Some of the beds have a scattering of dark, angular volcanic fragments which gave this quarry its old name of 'Rainspot' Quarry. Note that a recent (June 1980) collapse of underground workings has left parts of this quarry in a highly dangerous condition. This underlines the inadvisability of venturing underground into any old mine workings in the area.

Drive back to Keswick along the eastern side of Derwentwater under the imposing crags of Borrowdale Volcanics until, some 2 km from Keswick, the road passes onto Skiddaw Slates and the valley widens out into broad plain dotted with drumlins. On the right, immediately before the first roundabout, is the tree-covered knoll of Castle Head (270227), which should be visited after parking down by the lakeside (266229). Walk back through Cockshot Wood and climb to the top of Castle Head, a roughly circular outcrop of dark, fine-grained dolerite with glittering **augite** crystals. Castle Head is regarded as a volcanic plug filling a former vent cut through Skiddaw Slates, but from its small size and limited metamorphic effect on the surrounding Skiddaw Slates it can only have been a minor feeder vent of some larger volcano contributing to the Borrowdale Volcanic Group. This locality is a National Trust property and should not be disfigured by hammering.

The view from the top of Castle Head illustrates many of the features seen during the day's excursion and is a fitting finale to a trip through some of the finest scenery to be found anywhere in the world.

9 The Buttermere Valley

(1 day)

Purpose: To examine rocks of the Skiddaw Group exhibiting a variety of sedimentary structures, and folding, faulting and some mineralisation; the junction between the Skiddaw and Borrowdale Volcanic Groups; and in the latter group a number of rock types with faulting and mineralisation. The effects of glaciation are clearly seen.

The excursion comprises a number of fairly short but steep walks commencing from stopping places on the B5289 road running alongside Crummock and Buttermere. Some of the going is rough or boggy and exposed. Full fellwalking gear, including boots, should be worn.

OS maps: 1:25 000 Outdoor Leisure Map, The English Lakes, North-West Sheet (This shows details of footpaths and place names.)
1:50 000 Sheet 89
1:63 360 Lake District Tourist Map

There are refreshment and toilet facilities at Buttermere village, and coach parks at Buttermere and Gatesgarth.

GEOLOGICAL SETTING

The B5289 traverses rocks of the Skiddaw Group throughout the described route (Fig. 30). The altered mudstones, sandstones and **turbidites** of this Group have suffered severe deformation at more than one stage in their geological history and they show both small-scale and large-scale folding, as well as normal and thrust faulting. In the region of Grasmoor thermal metamorphism is evident in rocks forming the craggy front of the mountain (165204). The oldest rocks in the area belonging to the Skiddaw Group are the Hope Beck Slates found around Buttermere village (175170), followed by Loweswater Flags at Lanthwaite Gate (158210) and alongside Buttermere Lake. Kirk Stile Slates are to be found immediately below the volcanic rocks on the approaches to Honister Pass, whilst the youngest formation in the Skiddaw Group, the Latterbarrow Sandstone, is not seen in this area.

Rocks of the Borrowdale Volcanic Group surround the head of the valley in Warnscale Crags, on the north face of Hay Stacks (195133), the upper parts of Fleetwith Pike (205142) and the summit of Honister Pass where the pale green cleaved volcanic tuff is quarried and processed for roofing slate, wall cladding and building stone. The imposing ridge of High Crag (180140), High Stile (169147) and Red Pike (161154) is composed of andesite, some **dacite**, and tuff, resting on Skiddaw mudstones to the south of Buttermere, and intruded by granophyre from the vicinity of Sour Milk Gill (170159) northwestwards to Scale Force (150170).

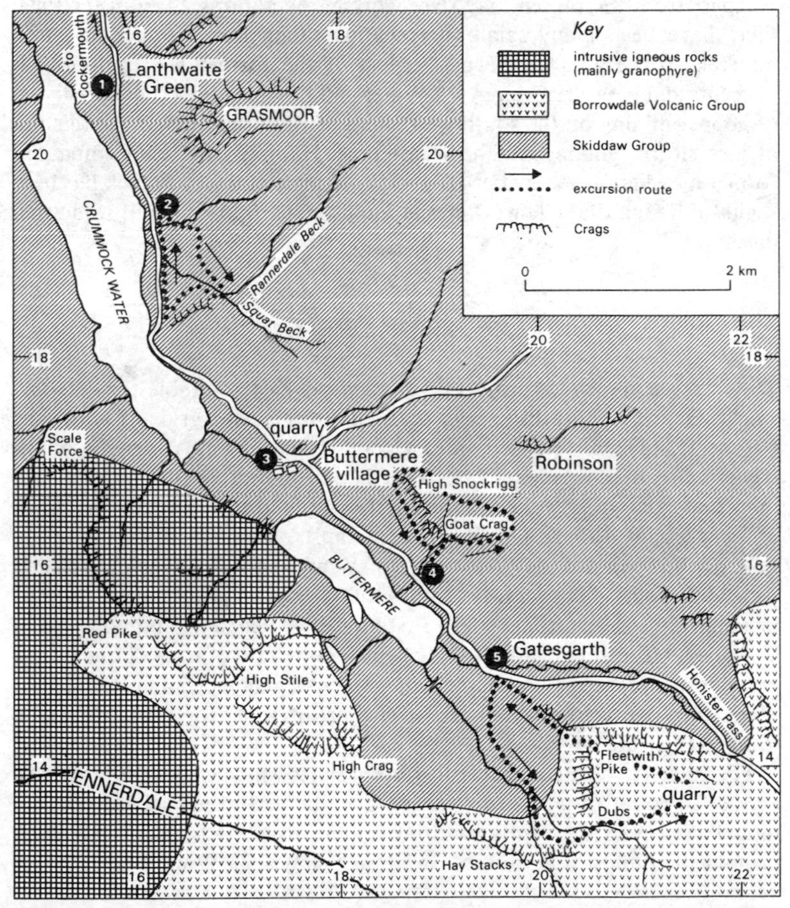

Figure 30 The Buttermere Valley.

Glacial and postglacial landforms are much in evidence. The north-east-facing combes (cirques) to the south of the lakes, Ling Comb (160160), Bleaberry Combe (165155) and Burtness Combe (175145), are striking in appearance. Only Bleaberry Combe contains a tarn. Warnscale Bottom is clearly a trough end, Fleetwith Edge (200146) is a fine *arête*, and Scale Force is one of many waterfalls in the area, tumbling from its hanging valley. Postglacial deposition, probably the result of torrential meltwaters from melting glaciers, loaded with glacial and solifluction debris, has divided Buttermere from Crummock Water, and deposited areas of flat, fertile alluvial land adjacent to the lakes at Rannerdale, Hassness and Gatesgarth.

Apart from the present-day slate-working operations at Honister Pass, there have been many small ventures in mining lead, copper and iron minerals in the area in the past. Most of the former lead mining was concentrated between Crummock Water and Loweswater. There are traces of copper mining on the south-west shore of Buttermere, and the remains of a small iron mining trial at Scale Force. Happily, the visual impact of former metal mining activities around Buttermere and Crummock has been slight, although the slate workings at Honister are impressive, if somewhat unsightly.

EXCURSION DETAILS

The first recommended stop is a roadside car park, suitable for cars and small coaches, a short distance to the south of the cattle grid at Lanthwaite Green (**1**) (158206). This provides a good vantage point for much of the scenery already described above. The shapely mountain of Mellbreak lies immediately across the lake to the south-west, and to the south the cirque of Ling Comb with Red Pike behind it can be seen, weather permitting. East of the road are the bold fronts of Whiteside and Grasmoor, clearly truncated spurs. The craggy nature of Grasmoor End is due to the metamorphosed flagstones and gritstones of which it is composed. The V-shaped, steep-sided gash of Gasgale Gill is worthy of inspection. It may be approached across the damp but gently sloping open fell east of the road, scree paths then leading well above stream level; the better path goes to the left of the stream.

The gill itself is cut along a fault zone in slates which crumble under frost action, so rounded landforms predominate here, and glacial features are muted. Some of the faults have been mineralised to a small extent. A prominent mining trial adit on the lower slopes of Whin Ben (164209) below Whiteside is worth investigating. Allow a ½–1 h at this location.

An interesting alternative to the above is the Rannerdale and Buttermere

Hause area. Cinderdale Common is the best car parking area **(2)** (162194). Walk along the track from the car park southeastwards into Rannerdale. In front is the impressive rock wall of Rannerdale Knotts in thermally metamorphosed Hope Beck Slate with its series of cliffs controlled by joints and bedding planes. Craggy outcrops immediately to the left of the path are good examples of *roches moutonnées*, ice-scraped knolls of bedrock. Notice how the little valley widens and flattens upwards to the south and south-east where the stream is patently underfit (too small for its present valley). After about 1 km from the start, cross the turbulent streams at the confluence of Rannerdale Beck and Squat Beck (168186) to follow the path below the rock wall and rejoin the road at Buttermere Hause. Across the lake can be seen the continuation of the hardened slate formation in the truncated spur of High Ling Crag on the flank of Mellbreak. At its foot is Low Ling Crag (156184), a textbook *roche moutonnée* linked by a tombolo, a spit of gravel, to the shore of Crummock Water. Allow about 1 h at this location.

Buttermere village provides a useful stopping place for refreshments. Cars and coaches may be parked at the village car park near the Fish Hotel **(3)** (174169).

The cascade of Sour Milk Gill on the fellside across the valley, tumbling out of the hidden combe where Bleaberry Tarn lies, dominates the view. Chapel Crags, above the combe, are composed of rocks of the Borrowdale Volcanic Group linking Red Pike and High Stile. To the east are rounded hillsides leading up to High Snockrigg and Robinson, formed of Loweswater Flags of the Skiddaw Group. By the roadside near the Old School Room (now occupied by a Mountain Rescue exhibition) and the Church, tight folding in the slates, with NE–SW Caledonoid trend can be traced. In the abandoned quarry just north of the village (173173) and on the fellside above, folding, cleavage, jointing, quartz **stringers** and sole marks show well. The folding is quite complex; it is represented diagram-

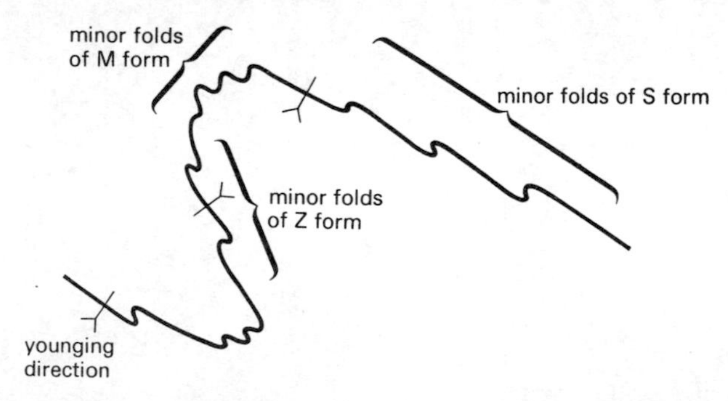

Figure 31 Fold form and relation in rocks of the Skiddaw Group at Buttermere.

Figure 32 Buttermere, location **4.** View of Goat Crag, looking north from Buttermere lakeshore, showing the faulted ravine of Hassnesshow Beck. (Photograph: J. D. Hinde.)

matically in Figure 31. Please do not hammer in the quarry or in the vicinity of the village.

The minor fold forms have resulted from unequal development of the two limbs. The minor folds are generally of the same shapes as the major folds produced by the same phase of deformation, but vary in form according to their position on a major fold structure. The main cleavage that has been induced in the shaly rocks of this area is of the pressure solution type. It is not uniform throughout the mass of rock but occurs in discrete planes along which the soluble quartz particles have been dissolved out, causing the insoluble micas to rotate and become aligned along cleavage planes. The effect of pressure solution is very intense in some places and has resulted in rock structures with the superficial appearance of thrust zones. There are probably two main periods of deformation represented hereabouts: open folding of middle Ordovician age, and intense folding with related cleavage associated with the **Caledonian** orogeny in late Silurian to early Devonian times. Allow about 1 h in Buttermere village.

Folded turbidites of the Loweswater Flags may best be seen in the vicinity of Goat Crag (190164; Fig. 32). There is limited space for car parking on the roadside at the S-bend near Hassness (**4**) (187159). A coach party should leave their vehicle at Buttermere and follow the track on the north shore of the lake before turning left (north) to rejoin the road near Hassnesshow Beck. Walk 400 m uphill alongside Hassnesshow Beck to a small water catchment dam at the foot of the crags. Exposures in both banks of the beck show superb turbidite and interturbidite sequences. Evidence from sole marks, **ripple drift lamination** and graded bedding may be used to determine whether the strata are inverted or not. The interactions of cleavage planes and folded bedding planes are well displayed (Fig. 33), as also are the effects of pressure solution. A generalised turbidite sequence is shown in Figure 34. It is thought that these rocks originated from a series of underwater slurry flows or turbidity currents rushing at high speeds down a continental slope, travelling far across and ultimately settling on the floor of an ocean. A precarious scramble into the deep ravine behind the water catchment pond will reveal clear turbidite structures in the water-worn rocks of the stream bed and in the weathered rocky sides. The ravine itself is clearly cut along a fault plane, showing a thick lens of crystalline quartz several metres long (Fig. 35). The turbidite sequence may be followed up a path on steep ground to the right (south-east) of the ravine, eventually bearing leftwards (northwestwards) over rough ground well above Goat Crag to a superb viewpoint on High Snockrigg (187168).

Descend via scree and rough fellside to Low Snockrigg and traverse below Goat Crag to Hassnesshow Beck. This ascent and descent on

Figure 33 Buttermere, location **4**. Skiddaw Group turbidite showing bedding and cleavage. Hassnesshow Beck (188161). (Photograph: J. D. Hinde.)

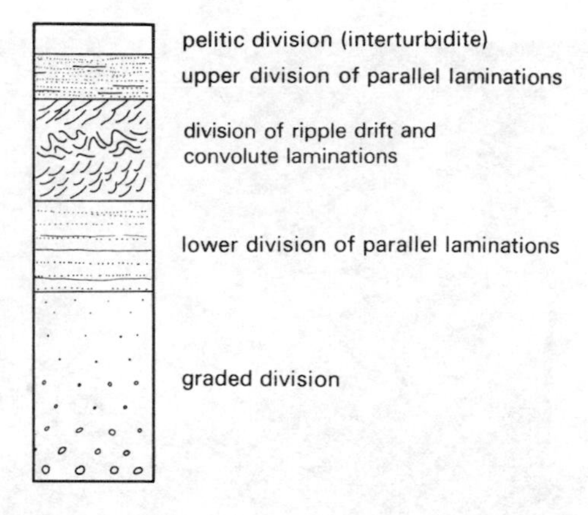

Figure 34 Turbidite divisions.

Snockrigg is only for *fit* parties in *good* weather. The alternative is to work leftwards (north) from the water catchment pond over the broken ground at the foot of Goat Crag looking at folding, cleavage, and in particular gathering evidence on the 'way up' of the strata. Allow between 2 and 4 h at this location.

Figure 35 Buttermere, location **4**. Crystalline quartz, with **slickensiding**, along a fault plane. Hassnesshow Beck, Goat Crag (190162). (Photograph: J. D. Hinde.)

Figure 36 Buttermere, location **5**. The Warnscale area. (Photograph: J. D. Hinde.)

Warnscale Bottom and Fleetwith Pike can best be approached from Gatesgarth. There is a car park and coach park on the roadside near the farm **(5)** (195150). Follow the path from the roadside above Gatesgarth Cottage southwards past the foot of Fleetwith. Once in Warnscale Bottom, follow Warnscale Beck up to a marked steepening of ground (Fig. 36). At the foot of the lower waterfall (201135), tightly packed folds in the less competent slate are followed abruptly by unfolded andesite. This marks the junction between the Skiddaw Group and the Borrowdale Volcanic Group, which can be seen to run along the fellside westwards below Warnscale Crags (195133), and out over the shoulder of Fleetwith Pike between Striddle and Raven Crags (202144). Walk up the lesser path on the south side of Warnscale Beck and rejoin the main path above the upper waterfall near Dubs Bottom (205134). Note the immense variation in the course of Warnscale Beck, from the broad flat floor of the trough end, the waterfalls controlled by differing hardnesses in beds of volcanic rocks, and finally the marshy corrie of Dubs Bottom (209133). Follow the track to Dubs Quarry.

In the vicinity of the climbing hut there is an abundance of **slate** waste, pale green cleaved volcanic tuff. The quarry is in part of a long outcrop of intensely stressed volcanic rock extending through Honister, Castle Crag and Quayfoot in Borrowdale (Excursion 8). Walk through a defile in the quarry waste eastwards towards Honister, and after 300–400 m bear left over the open fellside to some obviously fresh dumps of quarry waste above. This is Hopper Quarry (214137), currently being worked for slate, and showing a number of interesting features including a very prominent slickensided, mineralised fault plane, to the right of which is a clear section through the scoriaceous top of a lava flow which was interpenetrated with fine-grained volcanic tuff (Fig. 37). As the quarry is being worked sporadically, access may be restricted. Please heed the warning notices. Permission to enter should be sought in advance from The Buttermere and Westmoreland Green Slate Co. Ltd, Honister Hause, Borrowdale. From the quarry, head north-west for 1 km to Fleetwith Pike (205142), an unpleasant route in cloudy conditions, as the path is indistinct. There is a spectacular panorama of the flat-floored, straight-sided, glaciated trough of the Buttermere valley. Differences between the steep but rounded slopes formed from rocks of the Skiddaw Group on Robinson to the north, and the craggy north-east-facing rocks of the Borrowdale Volcanic Group on the High Stile ridge to the west are notable. Truncated spurs and the corrie of Burtness Combe show up well. Take the path down Fleetwith Edge, a stepped *arête* where tuffs form much steeper slopes than do the lavas. The northeastern side of the ridge, overlooking Honister Pass, is impressive. If you still have the energy, look for the Skiddaw–Borrowdale junction (202144) above Low Raven Crag where gently-dipping flow breccias are underlain by steeply-dipping mudstones. Finish walking down past the

Figure 37 The upper Honister Slate Quarry (214137). (Photograph: J. D. Hinde.)

white cross near the end of the ridge towards Gatesgarth. Allow between 3 and 4 h for this round.

10 St Bees Headland

(1 day or ½ day)

Purpose: Examination of the Carboniferous/New Red Sandstone unconformity and some of the rocks above, including Brockram, Magnesian Limestone, St Bees Shale and St Bees Sandstone, the last-mentioned exhibiting sedimentary structures. Specimens of gypsum and anhydrite may be collected, while a variety of igneous rocks and semi-precious stones found as beach pebbles have been derived from glacial drift, this forming low cliffs near St Bees.

Route A involves 9 km of walking, mainly on footpaths, with the return from St Bees to Whitehaven made by private or public transport (bus or train). Routes B1 (**1,2,3**), B2 (**4,5**) and B3 (**6,7,8,9**) may be undertaken as a series of shorter (½ day) excursions. Scrambling on cliffs is involved, so boots and adequate clothing should be worn, with the wearing of safety helmets advised in the vicinity of steep faces at **1,3,4,5,7,8**.

OS maps: 1:25 000 Sheet NX 91 shows all required details
1:50 000 Sheet 89 shows access roads and footpaths
IGS maps: 1:50 000 Sheet 28 (Whitehaven); solid *or* drift edition

The starting point is near Whitehaven (see excursion account for details). There is an extensive public car park with toilets and refreshment facilities at St Bees beach (Seacote), while cars may be parked on roadsides at Kells and Sandwith.

GEOLOGICAL SETTING

The St Bees Headland forms a distinctive promontory some 100 m high and 14 km^2 in area lying to the west of a broad, prominent north–south-trending glacial meltwater channel that runs from Whitehaven to St Bees. The northern part of this small upland area, occupied by the Kells and Woodhouse districts of Whitehaven, is underlain by Upper Carboniferous Whitehaven Sandstone and Coal Measures. Of the once numerous coal mines in the area only one, the Haig Colliery (967176), remains operative, its workings extending several kilometres westwards under the Irish Sea. A large quarry situated in the hillside behind the town cemetery (973165) worked Coal Measures shales for brick clay until a few years ago, but is now derelict.

The western and southern parts of the headland form the highest ground in the area, being underlain by strongly jointed red St Bees Sandstone dipping gently WSW. The St Bees Sandstone is the highest formation in the New Red Sandstone in the area, and is underlain successively by St Bees Shale which contains evaporite (gypsum and **anhydrite**) horizons, Magnesian Limestone and Brockram.

The basal Brockram, a breccia of Permian age, hereabouts is very thin (*c*. 2 m) and is thought to form the edge of a series of ancient desert scree fans which were banked up against a mountain massif on the site of the present Lake District in Permian times. The New Red Sandstone formations rest uncomformably on previously tilted and eroded Carboniferous strata, and this unconformity is well seen at Barrowmouth (960161).

During the past 2 Ma the area has been invaded by ice streams emanating from icefields situated on the Lake District and the Scottish Southern Uplands. Rapid downwasting of the Lake District ice when the Irish Sea basin was still choked with Scottish ice led to the formation of a complex series of meltwater channels trending north–south down the Cumbrian coastal strip. The former presence of Scottish ice in the area is inferred from ‘exotic’ boulders and pebbles of Scottish rocks (e.g. Criffel Granite) in till deposits. Pebbles derived from these tills are particularly well seen at Fleswick Bay (945134) and along St Bees beach (around 964114). The Golf Course cliffs backing the beach at St Bees show a fine mixture of outwash sands and gravels and coarse blocky moraine, reflecting very rapid oscillation in the local glacial regime.

EXCURSION DETAILS

Routes A and B1 start from Kells (967165) on the hill 1.5 km SSW of Whitehaven. Vehicles may be parked on the roadside or (by arrangement) on one of the large open car parks belonging to Messrs Albright and Wilson Ltd, Marchon Division. Make for the private road, permissible as a right of way on foot, which skirts the northern and western boundary of the Marchon Factory, cross the railway line and head for the rising ground between an industrial tip and the factory fence, where there is a small quarry near the ruins of Lingydale Farm (**1**) (963160). The quarry has been partly filled with soil and rubbish, but there is an abundance of anhydrite and gypsum specimens discarded from the Marchon anhydrite drift mine opened up nearby to provide the raw material for sulphuric acid manufacture. The mine has now been ‘mothballed’ following a switch of the sulphuric acid plant from local anhydrite to imported sulphur as its raw material. The quarry shows a number of interesting features. There is an

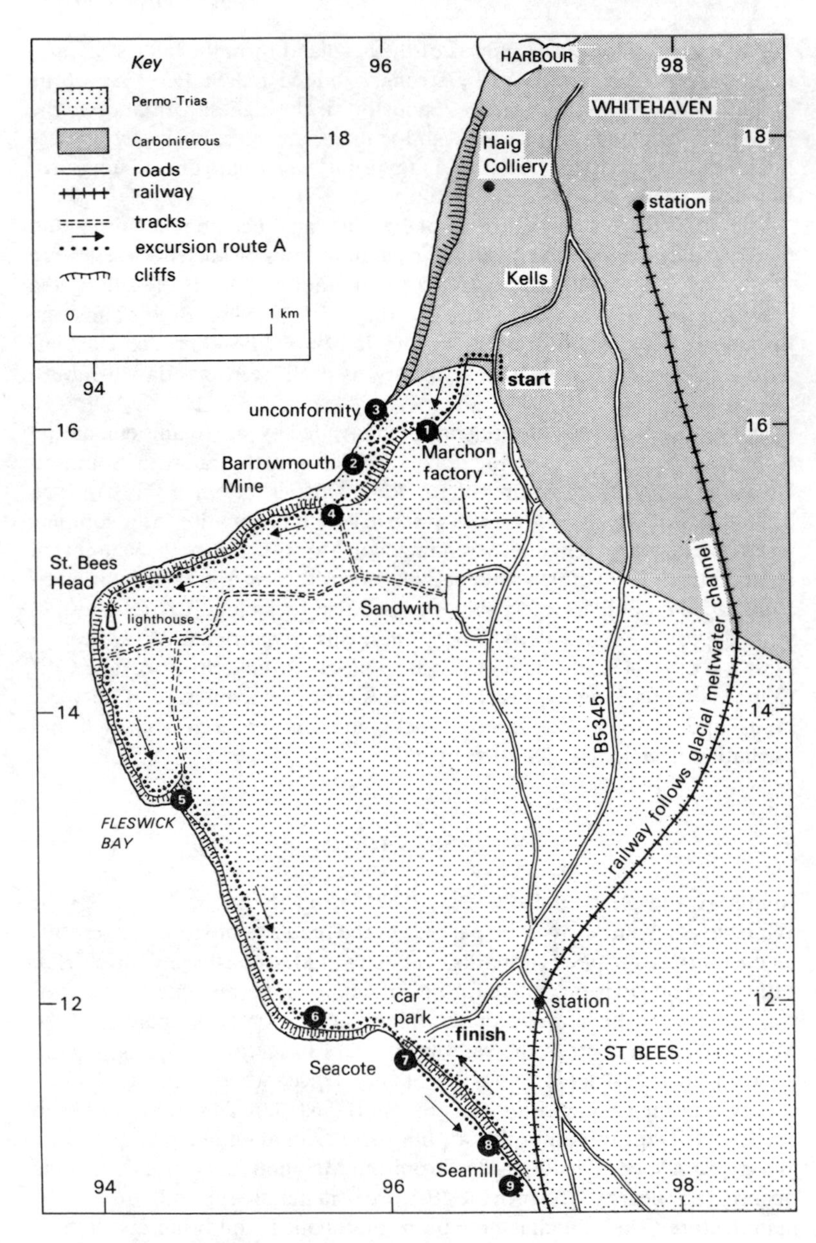

Figure 38 Sketch map of the St Bees Headland.

upward gradation from shale, which was quarried for use in the anhydrite–sulphuric acid process, to sandstone. There is no distinct junction between this St Bees Shale and St Bees Sandstone, however. Fallen blocks of the sandstone give interesting clues to the origins of the rock; ripple marks, **load casts** and desiccation cracks being particularly noticeable. The predominantly mahogany red colour of the shale and sandstone is broken here and there by green layers and blotches, presumably where the red iron(III) oxide staining has been chemically reduced to the iron(II) state. At the factory end of the quarry the effect of soil creep on the dip of the beds can be clearly seen, a caution to geologists taking dip measurements on steeply sloping hillsides. It is worth recording the general dip of the beds in the quarry and relating it to observations made later.

From the quarry walk westwards along the level track to a point where a former mineral incline led down to the Barrowmouth Alabaster Mine. The ruined mine buildings can be seen a few metres above sea level at the foot of the incline. Notice that the incline, which was once presumably straight to allow tubs to be hauled up it, is now markedly kinked (Fig. 39). The reason for this becomes apparent when one considers the shaly nature of the underlying rock and looks along the cliff slopes at several recent and ancient landslip scars. Note the dip of the sandstone/shale boundary southwestwards and the manner in which the weathered shale produces

Figure 39 St Bees Headland, locations **2** and **3**. The Barrowmouth area. (Photograph: T. Shipp.)

steep slopes whilst the well jointed sandstone forms vertical cliffs. Walk down the incline towards the ruined mine buildings (**2**) (957157). The sites of the mine adits are marked by springs behind the ruins. Dip measurements on the old mine buildings will show them to be tilted up to 15° back towards the cliffside due to the rotational effect of land slippage hereabouts. There is a small dump of mine waste, mostly anhydrite. The mine, which was worked in the middle of the last century for alabaster and plaster production, failed because of the diminishing gypsum and increasing anhydrite encountered on penetrating the St Bees Shale.

From the mine walk 300 m NNE and descend the slippery clay to the foreshore, which here is largely composed of boulders of St Bees Sandstone. The wave-washed rock bench and low cliff contain the unconformable junction between the Upper Carboniferous and New Red deposits (**3**) (960161). Working upwards in the succession note the base of purplish Whitehaven Sandstone, markedly false-bedded, with a weathered and irregularly eroded upper surface.

Approximately 2 m of Brockram rests on the Whitehaven Sandstone and in places penetrates into it down opened joints. The composition of the Brockram should be investigated. It seems to consist mainly of sandstone (probably Carboniferous), limestone (certainly Carboniferous as indicated by fossil corals and brachiopods) and occasional weathered tuff and lava fragments (probably Borrowdale Volcanic Group). The limestone fragments have been heavily dolomitised, many of them showing hollow centres. Overlying the Brockram is 5 m of buff-coloured Magnesian Limestone, deposited from the highly saline Zechstein Sea in Permian times. It contains a sparse fossil fauna of the highly stunted bivalve *Schizodus* sp. The Magnesian Limestone passes gradually upwards into buff-coloured shale and the red St Bees Shale already encountered in the early part of the excursion. Head back to the former mineral incline and climb it to where a well trodden path skirts below outcrops of St Bees Sandstone. Route B1 from here returns to Kells.

Route A follows the path which rises gently in a southwesterly direction. There are excellent views of the coastline westwards and northeastwards. The St Bees Sandstone is markedly false bedded, and the undersides of some of the projecting beds in the upper cliff face show excellent examples of load casts. Route B2, which starts from Sandwith village (965147), joins Route A at the cottages alongside the disused Birkham's Quarry (**4**) (956154). From the cottages head west along the cliff-top footpath which winds in and out along a series of old cliff-edge quarries.

St Bees Sandstone from these quarries has been used as a beautiful and durable building stone throughout Cumbria. It was even exported to the American Colonies as ballast in eighteenth-century sailing ships out of Whitehaven. Whilst the rock is generally sound hereabouts, the path skirts very close to some sheer drops, so care should be taken, particularly in wet and windy conditions. Good cross-sections of sedimentary structures such as ripple marks and cross-bedding may be found in these cliff-top quarries. Carry on towards the St Bees Head Lighthouse, in the vicinity of which a large pillar of sandstone has leaned away from the cliff edge. The bedding planes and joints in the sandstone, etched out by the weather, form ideal nesting sites for a variety of seabirds. Two separate areas of cliff adjacent to the footpath are now under the protection of the Royal Society for the Protection of Birds, and should on no account be interfered with.

Follow the field path downhill to Fleswick **(5)** (945134), a steeply cut southwest-facing valley, presumably incised along a fault or major joint plane. The beach at Fleswick Bay is composed of a relatively thin veneer of gravel containing beautifully rounded small pebbles of many varieties of quartz and igneous rock. Semi-precious stones such as agate, carnelian and jasper may be found. These pebbles form the residue of fragments of hard material from glacial drift which has been eroded by the sea. They have been rounded and polished by abrasion against each other and against the St Bees Sandstone. The effects of this on the latter can easily be seen at the undercut base of the cliffs, and at low tide when the fluted and scoured wave-cut platform is laid bare.

On a falling tide it is possible to scramble along the foot of the cliffs towards St Bees village, but the going, although interesting from the point of view of structural and erosional features in the sandstone, is very rough. It is preferable to walk back up the Fleswick valley. Route B2 continues due north over fields for 1 km before turning right along the road past Tarnflat Hall to Sandwith village.

Route A turns immediately southeastwards along the cliff path over Tomlin and South Head to Pattering Holes **(6)** (953118). This appears to be a landslip area with aligned cliff-top craters marking opened joints in the underlying sandstone. From here there is a good view of the seaward end of the St Bees valley with its cliffs of glacial deposits.

Walk down to the promenade at St Bees (Seacote). Route B3 commences here. Continue along the concrete promenade to the beginning of the boulder clay cliffs adjacent to the upper car park. A traverse of the base of the cliffs from this point (962117) to Seamill (969109) will be rewarding. The astonishing variety of pebbles at the top of the beach reflects the movement of glacier ice from both the Lake District and southern Scotland during the Pleistocene. Granites, **granodiorites,** andesites, tuffs and greywackes, together with much vein quartz, are present. There is also a great deal of locally derived sandstone, bright red Triassic and duller brown Carboniferous. A few metres beyond the southern end of the concrete promenade is a huge boulder of green vesicular andesite **(7)** (962116), an erratic, ice-borne from the Borrowdale Volcanic Group.

The beach material has been largely derived from the cliffs adjoining the golf course. Active erosion of these is evident as there are landslip scars and mudflows to be seen. The composition of the cliffs is very variable and it is possible that material has been deposited directly from an ice front as moraine, and from meltwater as irregularly bedded sand and gravel. At one point a thin peat bed above moraine and below sand high in the cliff face **(8)** (966112) indicates an amelioration of glacial climate. At the southern (Seamill) end of the cliffs a search at low tide will reveal tree stumps and roots in grey clay **(9)** (968108), a submerged forest. From Seamill a path along the cliff edge may be taken back to the Seacote car park, or a road under the railway line may be used for access to St Bees village for bus stop or railway station.

11 The Granite of lower Eskdale

(1 day)

Purpose: Examination of exposures of the Eskdale Granite, intrusive into rocks of the Borrowdale Volcanic Group, including intrusive contacts and remnants of the roof of the intrusion. The spoil heaps of a former haematite mine can be searched and glacial features seen, including a subglacial meltwater channel.

The excursion has been divided into distinct parts, each starting from a station on the Ravenglass and Eskdale Railway, known locally as the 'Ratty' and familiar to railway enthusiasts. This was opened in 1875 for the transport of iron ore from the Nab Gill mine at Boot, and later carried granite from Beckfoot Quarry. It is now entirely devoted to passenger traffic, and details can be obtained from: The Ravenglass and Eskdale Railway Co. Ltd, Ravenglass, Cumbria CA18 1SW. Walking distances should be judged from routes given in each part, and fellwalking gear worn where appropriate. Prior permission is needed to visit Beckfoot Quarry (see p. 102).

OS maps: 1:25 000 Outdoor Leisure Map, The English Lakes, South West Sheet (gives all required details and localities)
1:50 000 Sheet 96 (**1,2,4**)
1:50 000 Sheet 89 (other localities)
1:63 360 Lake District Tourist Map (all localities)

IGS maps: 1:50 000 Sheet 37 and 47 (one sheet), Gosforth and Bootle; solid *or* drift edition
1:63 360 Sheet 37, Gosforth

The initial excursion starts at Ravenglass station (085965), others at stations along the line; some can be omitted if advisable, or cars may be used. There is a good rail service from Easter to September, but the timetable should be consulted in advance. Toilets, shops, parking and an information service are found in Ravenglass, and toilets at Dalegarth (174007).

GEOLOGICAL SETTING

Two major events greatly affected the geology of the area.

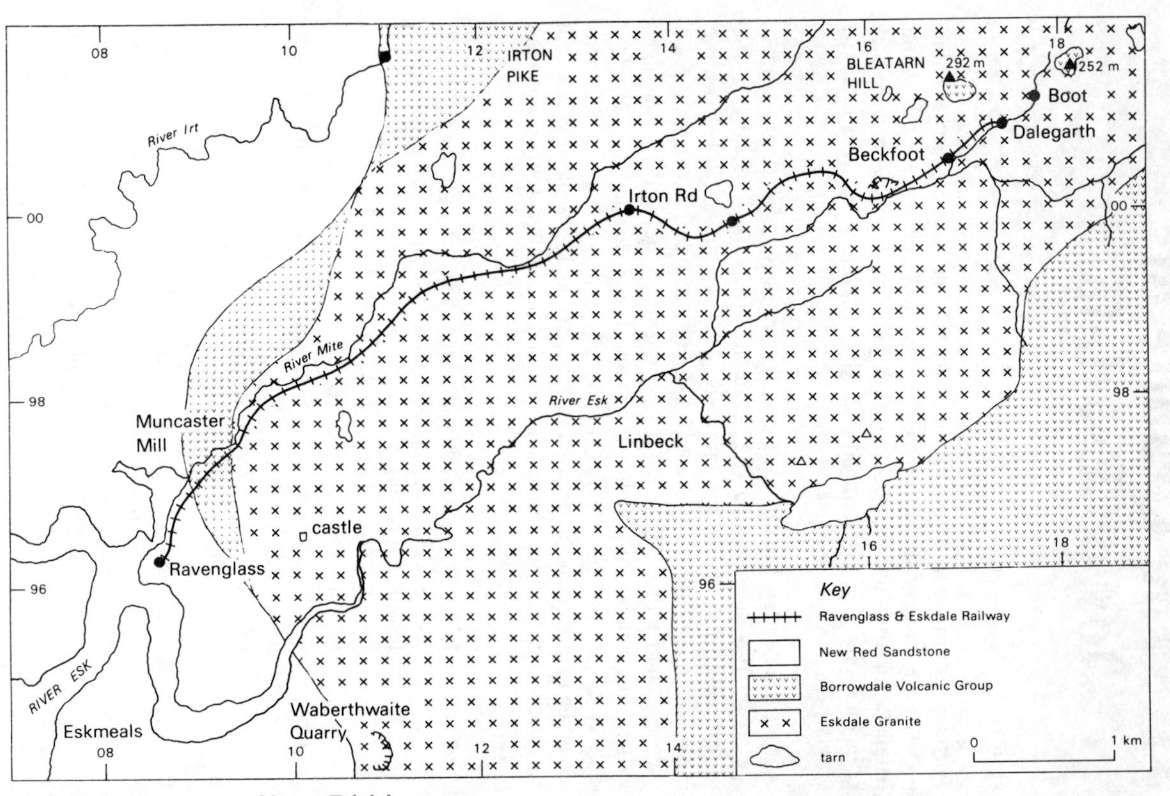

Figure 40 Sketch map of lower Eskdale.

Intrusion of the granite

A large granite stock, arising from a hidden igneous batholith under central Lakeland, was intruded into the andesites of the Borrowdale Volcanic Group (BVG) during the Caledonian orogeny. This is known as the Eskdale Granite. To the north it is overlain and partly hidden by the Buttermere and Ennerdale Granophyre, the junction with which is thought to occur at Easthwaite Farm (137035) near Wastwater. To the south it is intrusive into the Skiddaw Slates east of Bootle (115885). In the region of lower Eskdale its eastern flank extends beyond Boot to Wha House bridge (204009) where the junction with the BVG can be seen. The western limit is covered unconformably by the Triassic St Bees Sandstone, but since the latter lies below glacial drift, the true margin of the granite can only be postulated as an outward curve between Irton Pike (121015) and Muncaster Mill (096977) with a continuation due south. The granite boundary is outlined on Figure 40 and is based on the best available evidence. Gravity measurements indicate a steeply dipping western margin (Bott 1974).

The percentage of silica in the granite increases towards the margin of the intrusion and variation in composition occasionally occurs as at Linbeck (145979) where a so-called 'green' variety is found (**4**).

Furthermore, marginal intrusions associated with the granite mass are found at Irton Pike (**3**) (121015) and Waberthwaite Quarry (112944) consisting of **diorite** and granodiorite respectively. Discolouration often occurs in the form of red stains of haematite, seeping from the vertical cracks and faults in the rock. The faults in this region of the Lake District have a north–south trend as also have the veins of iron ore on both sides of the River Esk.

Evidence that the granite is intrusive into the BVG andesites is provided by the xenoliths of andesite at Waberthwaite Quarry and on the south-eastern flank, also by hardened andesite often darkened and discoloured by heat within the metamorphic aureole. Residual andesite caps the granite on the summits of Bleatarn Hill (168013) and Great Barrow (185016).

Glaciation

Ice sculptured this U-section valley of the Esk with its rounded crags, hanging valleys and *roches moutonnées*. An ice sheet advanced from the north-east and inundated the whole of Miterdale (Fig. 40), from which it was diverted to the south on meeting the ice mass of the Irish Sea and eventually covered lower Eskdale. A central granite boss beneath the glacier divided the two submerged valleys, and five meltwater channels were eroded from north-west to south-east across this obstruction below the ice.

With deglaciation the ice front retreated from the ridge of Muncaster Fell leaving the meltwater channels as depressions. The glacier was responsible for the drift material and rich boulder clay which produced the good soil of the valley and coastal strip; it also carried erratics of Eskdale Granite, some of which have been identified along the western flank of England and Wales far to the south.

EXCURSION DETAILS

Location 1: The Ravenglass Estuary (Fig. 41)

At the south end of Ravenglass main street, the view from the beach extends across the estuary and reveals at low tide a large tract of silt and sand, deposited by the rivers Irt, Esk and Mite. There is evidence from parallel shingle ridges on the Eskmeals Nature Reserve (082945) that the River Esk has been diverted northwards from its original southwesterly course. The end of the Nature Reserve can be seen from this vantage point and studies on the tallest sand dune have shown a northerly drift at a rate of 2.5 m per annum.

If the tide is out, proceed to the left alongside the sea wall for 300 m, where specimens of the Eskdale granite and andesites can be obtained from pebbles which have been deposited by the River Esk. These are erratics, released by sea erosion from the boulder clay and washed up by tidal action.

Location 2: Meltwater channel on Muncaster Fell (Fig. 41)

From Ravenglass railway station, walk up the hill to the war memorial and turn right along the A595, passing the car park, then continue 1 km to a sharp bend of the road where two paths emerge on the left side. The left-hand path goes through a gate and is signposted ‘Miteside’. Follow this path into Brankenwall Plantation. This wooded dell was a meltwater channel, one of five which cross Muncaster Fell from NNW to SSE; and with the stream below and almost vertical granite walls on the far side, it is easy to imagine how a sub-glacial torrent cut into the granite beneath the thick ice cap.

Proof of the direction of flow can be found from the glacial deposits which fan outwards below the terrace of Muncaster Castle. (The Castle is open daily except on I ridays.) Continue along the path until it reaches open ground with the River Mite in view. At this point the path (a) bends to the left and drops down to the main road just beyond Muncaster Mill from which a train may be caught back to Ravenglass.

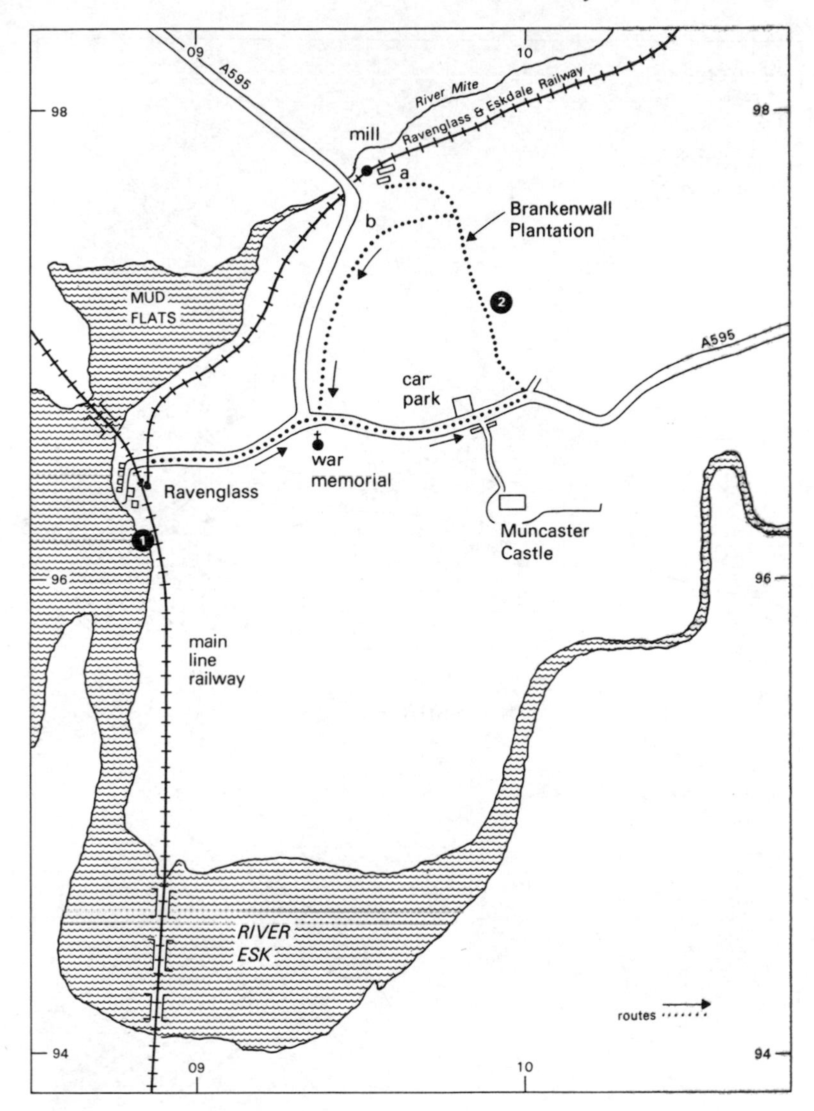

Figure 41 Locations **1** and **2** near Ravenglass.

As an alternative, an easier path (b) can be followed which circles alongside the main road and emerges above the war memorial; this route passes over flow-brecciated grey andesite showing little visible evidence of metamorphism though at the margin of the granite.

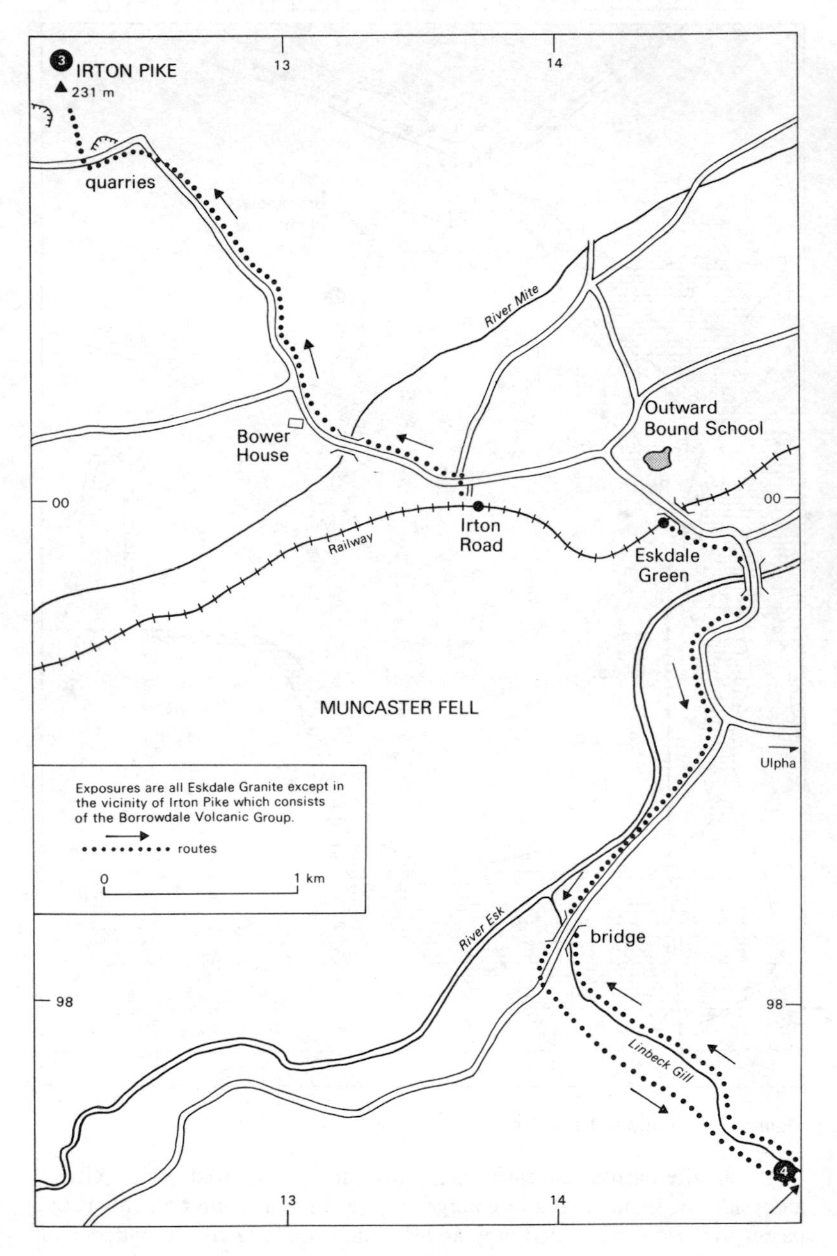

Figure 42 The Granite of lower Eskdale. Irton Pike (**3**) and Linbeck Gill (**4**).

Location 3: The granite boundary on Irton Pike (Fig. 42)

On leaving Irton Road station (137999) turn left along the main road to the Bower House Hotel and continue up the hill, avoiding the branch road on the left to Holmrook. Irton Pike, 231 m high, stands out ahead as a forested peak.

The first site to be visited is a small quarry behind a car park on the right, where Eskdale Granite is seen at the margin of the intrusion. Exploration amongst the undergrowth around this site (122013) reveals small exposures, some of which show pink granite intermixed with grey andesite, the latter having been heated and hardened by contact with the granite, thus demonstrating the close proximity of the two rocks at this location. Continue along the road, where a little further on an opening occurs on the right and a route can be followed up a firebreak, through the trees to the summit.

Near the start of this route a well defined track on the left leads to another quarry (120013) in a grey rock; this is diorite, containing quartz, pyroxene, hornblende and plagioclase. Continue to the summit (121015) where exposures of bluish andesite can be seen, some of which is porphyritic with white phenocrysts.

From the summit, Mecklin Park to the north-east marks the theoretical granite contact with the **BVG**, while the Eskdale Granite is thought to meet the Buttermere and Ennerdale Granophyre at the foot of Wastwater (137035). Return downhill to the road and cross it to the entrance gate of Parkgate forest, where just inside, a wide assortment of varieties of Eskdale Granite can be found. Return to Irton Road station.

Location 4: Granite variation at Linbeck (Fig. 42)

From Eskdale Green Station (146998) turn right to the King George Hotel (148998) and again right along the Ulpha road; 0.5 km beyond the bridge turn right to the exit of Linbeck Gill at another bridge (140982). Enter the forest to the left just beyond this bridge and follow the path alongside the gill until it emerges on the open fell, where approximately 800 m further on some outcrops can be seen (145979). These are composed of the so-called 'green granite' which is in truth light grey in colour and shows severe weathering. This granite has a lower silica content than the normal Eskdale Granite but a higher percentage of iron, with biotite present. The return can be made the same way, or if the stream is low it can be crossed and the gill followed below the Eskdale Granite crags back to the bridge.

Devoke Water at the source of Linbeck is worth mentioning, although the 'Ratty' service scarcely allows time for a visit. Because of rough terrain and bogs, it is best approached along a track from the Ulpha road (171977)

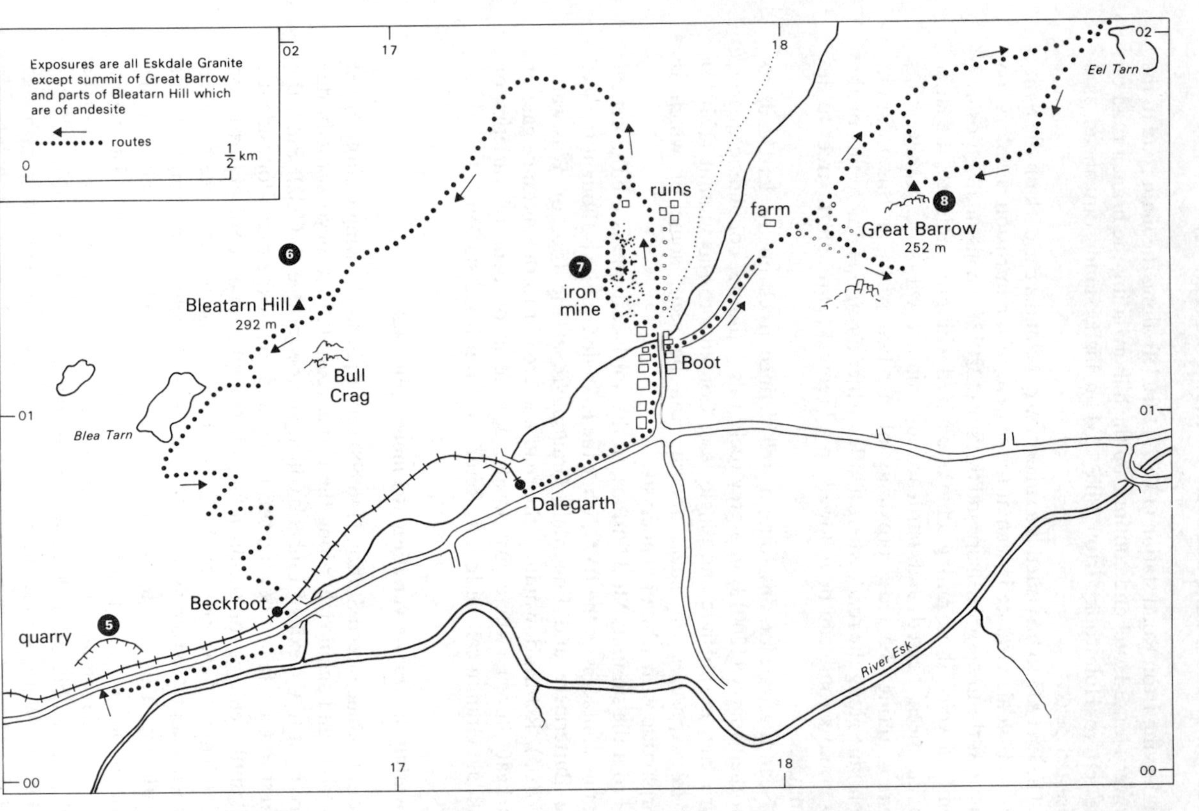

Figure 43 The area around Boot in mid-Eskdale.

Figure 44 The Granite of lower Eskdale. Beckfoot Quarry (**5**) (164003). (Photograph: F. Jones.)

and lies along a west–east fault plane. The main features are related to the boundary of the granite with the andesite and include displaced granite to the south (andesite dipping below the granite); while on Water Crag (153973) greisen and quartz-**sericite** are in evidence. The top of this crag still retains a small cap of hornfelsed Skiddaw Slate.

Location 5: The Eskdale Granite at Beckfoot (Figs 43 & 44)

The quarry (164003) is situated alongside the road, between the railway stations of Beckfoot and Eskdale Green. The entrance involves crossing the railway line to the gate, therefore care must be taken to ensure that no trains are in the vicinity. Permission to enter the quarry is required from Amey Roadstone Corporation Ltd, Threlkeld Quarry, Threlkeld, Keswick, enclosing a pre-paid envelope. An indemnity form must be signed. Protective headgear is advised, and the floor of the quarry is very marshy and overgrown and should be avoided. The quarry was opened up in 1901 and yielded between 38 000 and 57 000 tons of granite per year during the 1930s. In 1949 a single blast brought down over 50 000 tons of granite and yielded an average of 17 500 tons per year during the next three years. The quarry was closed down in 1953.

Eskdale granite is coarse-grained and pink in colour. Parts of the quarry face are stained red from haematite deposits, and drilled holes for shot blasting can be seen to the east. A search amongst the blocks at the base may reveal black, blue or brown tourmaline, wafer thin in nests or groups and usually associated with the joint planes.

As this area has been recently acquired by the National Trust, and the quarry is in active use for geological and biological research, large parties should avoid it, and on no account should hammers be used.

Locations 6, 7 and 8: Boot (Fig. 43)

From Dalegarth terminus turn left at the main road and continue eastwards for 400 m to the crossroads where you turn left to Boot.

Location 6: The roof of the granite at Bleatarn Hill (Fig. 43)

Cross the bridge at Boot (176012), pass through the gate and continue upslope on the left side of the stone wall (neglecting the gate on the right). The lane eventually passes between stone walls to three ruined buildings and, after leaving the fenced shafts of the iron ore mine, it turns first west, then south, and leads to a gully below Bleatarn Hill (169013). The summit of the hill consists of weathered granite which still retains the form of the upper surface of the intrusion. A search amongst the adjacent crags may be worth while to locate the remnants of Borrowdale Volcanic rocks, originally identified in this area by A. B. Dwerryhouse in 1909. The view from near the summit encompasses the whole of lower Eskdale. The return can be made via Blea Tarn, following a downhill zigzag path to Beckfoot station.

Location 7: The Nab Gill Iron Ore Mine at Boot (Fig. 43)

Cross over the bridge at the northern end of Boot village (176012) and pass through the gate immediately in front. The mine stands out as a red scar on the fell-side. The Nab Gill Haematite Mine (174014) was formerly owned by the Egremont Estate Company but is now the property of the National Trust. The mine was opened up in 1875, produced a peak output during 1880–81 but was forced to close down in 1913 when the bottom level flooded. The 'Ratty' railway was originally built to carry the ore to the coast from an assembly point alongside the now derelict headquarters. An average of 8000 tons a year was removed from five levels, the best quality being at the fell top. Other mines at Christcliffe (186012) and Gill Force (180000) operated at the same time but were in no way comparable in output to Nab Gill.

The haematite deposits follow NW–SE faults and joints in the granite on both sides of the River Esk, and are thought to have arisen by permeation from a higher sequence of rock. Some geologists regard the Permo-Triassic sandstone covering, now eroded away, to be the source. Some of the adits are now filled in, but the spoil heaps remain and are worth sorting for good specimens of haematite and the bright **'kidney ore'**. However, three shafts can still be seen behind the protective fencing at the top of the quarry.

Location 8: The cap of andesite on Great Barrow (Figs 43 & 45)

At the last white house on the right, prior to the bridge at Boot (176012), a lane leads off to the waterfalls. It then ascends to a gated farm entrance. Just before the farm gate is reached a wooden gate on the right gives access to a walled lane. On the right-hand side a little further on, a marshy ascent can be made between widely spaced walls to the depression between Little Barrow (183014) and Great Barrow (184016). Exploration of this region is rewarding. Little Barrow consists entirely of pink Eskdale Granite, but Great Barrow has on its summit a clearly defined layer of grey andesite to a thickness of 10 m which is completely isolated from the granite mass on all sides. A good view of the glaciated features of upper Eskdale is obtained with Scafell to the north-east. On returning from the depression between the two crags and to the walled lane, the route is resumed towards the north-east until Eel Tarn (189019) is reached on the right-hand side, where

Figure 45 The Granite of lower Eskdale. Andesite capping of Eskdale Granite, Great Barrow **(8)** (184016). (Photograph: F. Jones.)

the so called 'green' granite can be found (p. 95). The route is retraced back to Boot.

This excursion may be extended to the eastern boundary of the granite, as follows, although careful planning and attention to train times is required for this additional exploration which involves fell walking in difficult terrain. The use of map and compass is advised. In bad weather this section should be avoided.

Approximately 1.5 km from Eel Tarn, along a poorly defined path to the north-east, lies Stony Tarn (200025) and the eastern boundary of the granite. In order to find good exposures of the granite/andesite contact zone it is necessary to proceed a further 1.5 km to the south-east on to the crags between Whin Crag (200023) and Goat Crag (203017).

Good exposures show sharp angular fragments of grey andesite bedded into a mass of pink granitoid rock. From Goat Crag, a steep, awkward descent southwestwards leads to the road in the vicinity of Eskdale Youth Hostel (196011) and the Woolpack Inn (190010).

12 The Appletreeworth area

(1 day)

Purpose: To examine a faulted sequence of Ordovician and Silurian sediments overlying the Borrowdale Volcanic Group. Fossils, including graptolites, and some minerals may be found.

There is some 5 km of walking, mainly on Forestry Commission roads and tracks; stout walking shoes or boots should be worn.

OS maps: 1:25 000 Outdoor Leisure Map, The English Lakes, South-west Sheet (shows more detail)
1:50 000 Sheet 96
1:63 360 Lake District Tourist Map

The excursion starts at Hawk Bridge (239920) where there are ample parking facilities for cars. There are toilets, cafés, shops and parking facilities for coaches at Coniston.

GEOLOGICAL SETTING

Appletreeworth Beck (250933) forms the eastern boundary of the Dunnerdale Fells which comprise over 2462 m of tuffs, andesites and rhyolites belonging to the Borrowdale Volcanic Group. The overall structure is a large north-east-plunging syncline of pre-Caradoc age, cut by numerous north–south trending wrench faults which displace the rocks up to 500 m. The volcanic rocks in the Appletreeworth area belong to two formations. Those forming The Hawk (241924) belong to the upper part of the Dunnerdale Tuffs and are andesites with thin beds of tuff at the top. These are superseded northwards by the Lickle Rhyolites, a sequence of rhyolites with bands of tuffs and andesites (see table on p. xi). The Coniston Limestone Group of Caradoc age lies unconformably on the volcanic rocks and is displaced by small dip faults, the sequence being repeated by reverse faulting at The Hawk (Fig. 47).

Geological maps of the area are in preparation, by the Institute of Geological Sciences, the only available work on the Borrowdale Volcanic

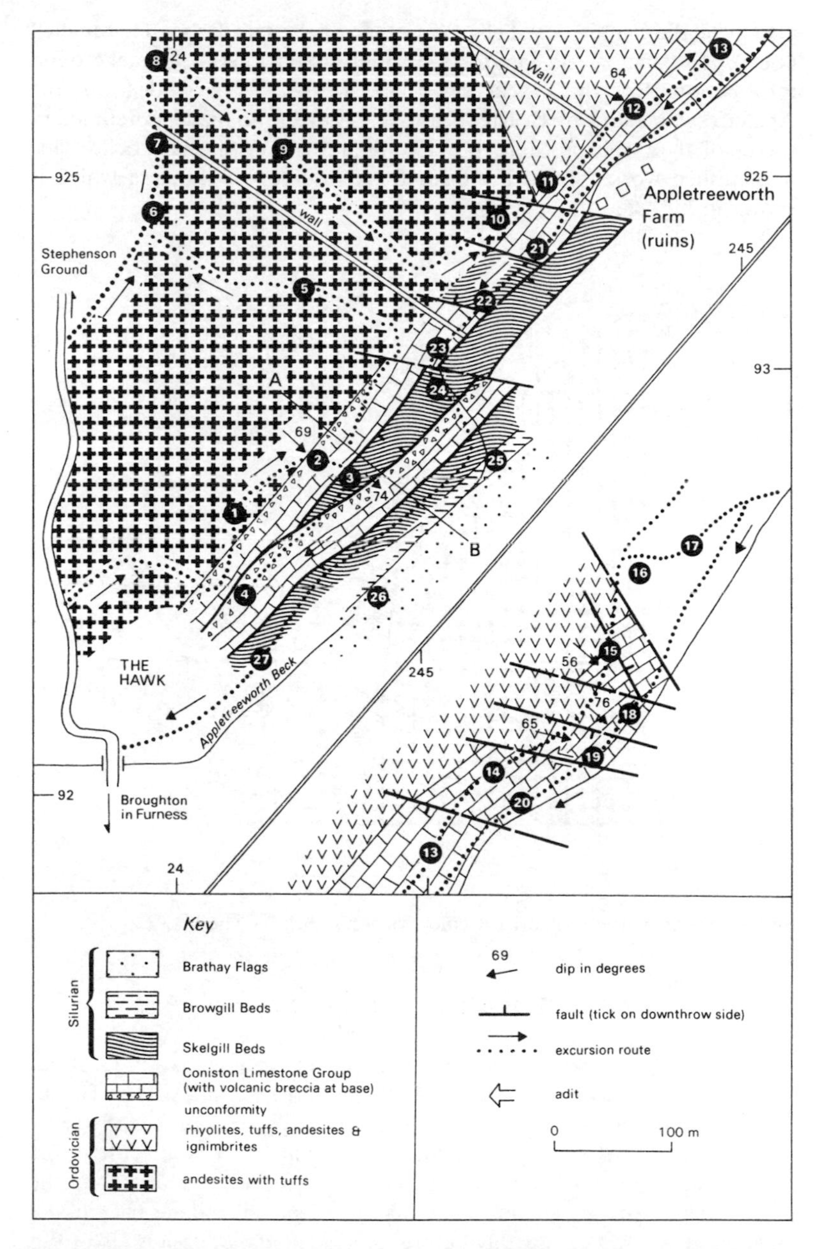

Figure 46 Geological map of the Appletreeworth area.

Group of the Dunnerdale Fells being that of the late Dr G. H. Mitchell (1956). His paper is very easy to read and it provides excellent background information for this excursion. The work was extended in the Appletreeworth area by C. Knipe and W. Grieve, who made a preliminary survey of the Coniston Limestone Group and Lower Silurian rocks. This unpublished work formed the basis for a resurvey of the area which is shown in Fig. 46.

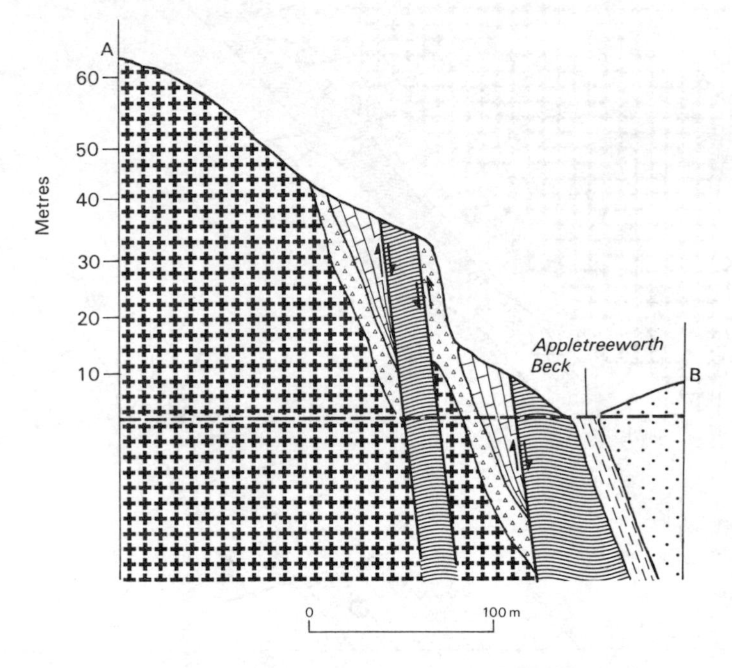

Figure 47 Appletreeworth area. Geological section across The Hawk.

EXCURSION DETAILS

From Hawk Bridge follow the steep road leading to Stephenson Ground, then after 200 m take the old track (239921) to the summit of The Hawk. On the edge of the clearing at the top can be seen small exposures of steeply dipping rocks of the Coniston Limestone Group. At **1** is an outcrop of amygdaloidal andesites containing vesicles filled with chlorite. In the nearby hollow are the remains of an old settlement which dates back to the early hundreds AD or possibly before. Amygdaloidal andesites also form the outcrop on the opposite side of the settlement and it appears that the contortions in the rock are accentuated by quartz veining. Moving down the eastern side of this outcrop small exposures of Coniston Limestone can be seen, the calcareous layers having been weathered into hollows. This is a good example of differential weathering and is very typical of the rock which is, at best, a very impure limestone. Follow the contour of the hill northwards passing hollows made by excavations into the limestone. Location **2** is opposite the corner of the wood in a particularly large circular depression. Here can be seen a breccia with andesitic clasts up to 7.0 cm long overlain by steeply dipping thin-bedded non-calcareous sandstones; this is the junction between the Borrowdale Volcanic Group and the Coniston Limestone Group (Fig. 48). Exposures like this are rare, so please do not hammer. Proceed now in a southeasterly direction and follow the edge of the wood to **3** where black shales of the Skelgill Beds are exposed, whilst a few metres further on another junction between volcanic breccia and the Coniston Limestone Group can be seen. At **4** volcanic breccia can be examined at its outcrop which trends in a southwesterly direction. The junction with the Coniston Limestone Group makes the base of a prominent escarpment nearby. So in a short distance there is a repetition of the volcanic breccia and Coniston Limestone Group and this can be explained by a strike fault throwing the Skelgill Beds down into juxtaposition with the Coniston Limestone, followed by a reverse fault

Figure 48 Appletreeworth, locality **2**. Andesitic breccia (left) overlain by basal sandstones of the Coniston Limestone Group (centre and right). (Photograph: D. Leviston.)

which brings the volcanic breccia into contact with the Skelgill Beds (Figs 46 & 47).

Return now to the exposures of Skelgill Beds at **3**, noticing the flat area they make, which was cultivated by the inhabitants of the settlement. Take the footpath northwards into the woods with more exposures of Skelgill Beds just off the track on the right. Follow the track round and proceed west to **5** where very coarse tuffs are exposed on both sides of the track. Proceed down to the main track to **6** where slightly amygdaloidal andesites can be seen beside the track. Within the woods a large outcrop can be seen to the left; weathering of the rock suggests that there is flow banding which has possibly been contorted, curving down almost vertically. Examination of a hand specimen reveals that the rock is composed of sub-rounded and angular clasts which have been heavily haematised.

The lime-green mineral is calcite, confirmed by applying a drop of cold dilute hydrochloric acid; it should effervesce. Evidently the presence of calcite and the haematisation result from a secondary mineralisation which appears very common in this area. One's first impression is that the rock is a tuff, since the clasts weather proud of the matrix. If it was a flow breccia the clasts would weather into hollows, leaving the matrix proud. However, the roundness of the clasts and absence of a definite matrix preclude this from being a tuff: in fact it is brecciated lava probably formed while the flow was still in a plastic state.

Figure 49 Appletreeworth, locality **7**. Fine tuff overlying an irregular surface (indicated by the pencil) of brecciated andesite. (Photograph: D. Leviston.)

Proceed up the track passing exposures of andesites, some containing phenocrysts of augite. Location **7** is reached just before the wall. Here another brecciated lava, but this time not haematised, is overlain on an irregular surface by a fine-grained tuff. Again, please do not hammer this exposure as there is not much of the tuff exposed. A photograph (Fig. 49) is provided so that you can identify these two units without the need to use your hammer. It is important to remember that the essence of good field geology is to use the hammer as little as possible, especially where exposures, such as this, are small. Let other people come and enjoy what you have just seen. The exposures on the far side of the wall are again of augite andesites. Augite is a common pyroxene, easily identified in hand specimen by its black colour and square or octagonal cross-section. Location **8** is a few metres up the track. Just before the bend is an exposure of green tuff containing black flake-like material which gives it a good cleavage. This is followed by a **welded tuff** and, right on the bend, a very coarse tuff.

Location **9** is in the woods on the left side of the track just before the junction with an older track. Here augite andesite is again exposed and if you have the patience to polish a sample you will have a beautiful specimen.

Proceed farther along the track, bearing left where the older track goes straight on. A good example of glacial drift (till) can be seen at **10**. Just before the junction with another track is **11**, where there is a large exposure of welded tuff or ignimbrite showing flow banding, small vesicles and amygdales. Continue now further down the track to the junction where a search in the undergrowth on the left will soon reveal pieces of slag. The smelters of the iron are not known but the slag has been dated as pre-Roman by the late Mr Morton of the Iron and Steel Federation Archeological Section. It could be that the inhabitants of the ancient settlement on The Hawk worked the small adit we shall visit at **19**. At this point the excursion may be shortened if desired by proceeding down the track to **21**.

Location **12** is about 100 m up the track, and here Coniston Limestone dips steeply towards the south-east. About 5 m up the bank is a small quarry. Ignore for the moment the right face and walk around to the left, where the uneven surface of a lava flow is overlain by a conglomerate with well rounded pebbles and boulders in a sandstone matrix, succeeded by the Coniston Limestone. Now examine the right face of the quarry where the beds are alternating bands of limestone lenses with thin sandstone partings. Ripple structures are common. Returning to the track further fossiliferous exposures can be examined, the first in particular having a bedding plane crowded with small brachiopods. PLEASE DO NOT HAMMER!

At **13**, 100 m further along, is a light coloured tough flint-like rock considered to be the volcanic 'ash' band, strictly a tuff band, which lies above the limestones (Fig. 52).

Continue 100 m up the track to **14** where dark cleaved shales which weather yellow are exposed just to the right of the track, inside the woods; these are the Mucronata beds (containing *Phacops mucronatus*) which are above the limestone. More limestone overlying rhyolite is exposed on the

Figure 50 Appletreeworth, locality **15**. Coniston Limestone lying unconformably on Lickle Rhyolites. The contact is shown by the broken line. (Photograph: D. Leviston.)

next bend, but the contact is not seen. Location **15** is a small quarry 200 m further on where limestone can be seen in contact with the rhyolite (Fig. 50). At the right of the quarry a fault plane is evident trending NW–SE, and this fault displaces the limestone to the north-west. The evidence of this is given by the small outcrops of limestone just on this corner. The limestone is not exposed at **16** so another fault must lie between these two localities. In the beck on the right of the track **(16)** dark, slightly calcareous shales are exposed, whilst downstream the sequence is ash followed by dark shales, culminating in banded black mudstones which are possibly Skelgill Beds.

Turn right at the junction with another main track, and in 70 m at **17** one would expect to find this sequence repeated, but this is not so. Instead Ashgill Shales are exposed and the slickensiding suggests that an east–west wrench fault separates the two localities. Fossil orthids have been found along this exposure, so it is always worthwhile searching for them. Notice how the shales have been cleaved to produce what is known as 'pencil' structure. From the evidence you have seen try and complete the geological map for these last two localities.

At **18** more limestone is exposed with the 'ash' band above the limestone cropping out a few metres further on. Evidently this has been faulted right out of step with the general line of the outcrop. Location **19** is about 70 m down the track and here a small adit has been driven along the fault line. In the nearby small tip samples of kidney ore and white **barytes**, with additions of **copper pyrites** and **azurite**, can be obtained together with smoky and clear quartz crystals. The working of this adit has not been recorded and it is certainly older than the last century, since even in those days nobody could remember its being worked. It is tempting to think that the settlement up on The Hawk, the pre-Roman slag and this adit are related in some way.

Further down the track Ashgill shales are exposed at **20**, and again fossils are to be found if some patience is exercised. Continuing along the path past the ruins of Appletreeworth Farm brings us back to just below the junction where the slag is found. Locations **21**, **22** and **23** are the three small quarries found further along the track. These present something of a problem, since in hand specimen the rocks found in them are very similar, being pink in colour with the texture of a tuff. Indeed in early mapping they were included as part of the 'ash' band that overlies the limestone. However, the discovery of fossils has proved that the rocks are sedimentary. Location **21** is in Ashgill Shales and if you are lucky you may find a complete trilobite. Unfortuntely the rock is so friable that there is virtually no chance of taking it home. Location **22** is in the Skelgill Beds and graptolites can be found, whilst **23** is back in the Ashgill Shales again. Clearly some process has been in operation to bleach or decarbonise these rocks.

In all these quarries pieces of rock can be found with fern-like patterns on them which are not fossils but dendrites of the black manganese mineral pyrolusite (MnO_2).

The reason for this repetition of Ashgill Shales is faulting, and there is much evidence of this at all three localities. Skelgill Beds crop out on the opposite bank of the stream.

At **24**, around the bend in the track, Skelgill Beds are exposed, with nodules of the iron sulphide marcasite (FeS_2) and graptolites. The whole exposure is covered with fault crossing fault to present a very complicated feature. This is not surprising when you realise that the exposure is bounded by a normal fault on the east and a reverse fault on the west. Continuing down the track volcanic breccia is overlain by limestone, with Skelgill Beds being brought down by a normal fault right on the bend.

From this bend descend to **25** in the stream bed where grey-green mudstones very typical of the Browgill Beds are exposed. Return to the track and continue down for about 200 m passing small exposures of volcanic 'ash' overlying limestone on the right. Location **26** is in the stream bed, where Brathay Flags containing good specimens of *Monograptus priodon* and *Cyrtograptus* sp. are exposed. Illustrations of these graptolites can be found in the Natural History Museum's publication on Palaeozoic fossils (see Plate 30).

Location **27** is again in the Skelgill Beds, with marcasite nodules common. From here it is only about 200 m to Hawk Bridge.

13 Greenscoe Quarry: an Ordovician volcanic vent

(1 day)

Purpose: Examination of agglomerates, tuffs and andesites in a volcanic neck formed in rocks of the Skiddaw Group, and of faulted limestones and mudstones of the Coniston Limestone Group.

Parts of the quarry are overgrown, particularly the upper quarry (7), and the going is rough, so that stout boots should be worn. The sides of the main quarry are subsiding in places, so safety helmets should be worn and great care taken. Greenscoe Quarry is private ground, and permission must be obtained from: Mr J. Taylor, Park Farm, Dalton-in-Furness; telephone no. Dalton-in-Furness (0229) 62587.

OS map: 1:50 000 Sheet 96
IGS map: 1:25 000 SD 27 Dalton-in-Furness with accompanying memoir, 'Geology and hematite deposits of South Cumbria'.

Parking along the grass verges at the entrance to the quarry is restricted to about six cars but there is a small parking area 60 m down the road.

GEOLOGICAL SETTING

Greenscoe Quarry is a volcanic neck intruded into black shales occupying the core of the northeasterly plunging Stewnor Anticline (Fig. 51). The black shales have yielded graptolites ranging probably from the zone of *Didymograptus hirundo* to the zone of *D. bifidus* or *D. murchisoni* indicating an age ranging from that of the upper Skiddaw Group in the shale quarries around Park Farm (218755) to that of the Eycott Group in the vicinity of Greenscoe Quarry (Rose & Dunham 1977, Wadge 1978). Since limestones of the Coniston Limestone Group (Fig. 51) rest **unconformably** on rocks of both Skiddaw and Eycott Group ages, and are unaffected by the volcanic activity within the neck, this places an upper limit on the age of the eruption. It must therefore be of late Eycott or

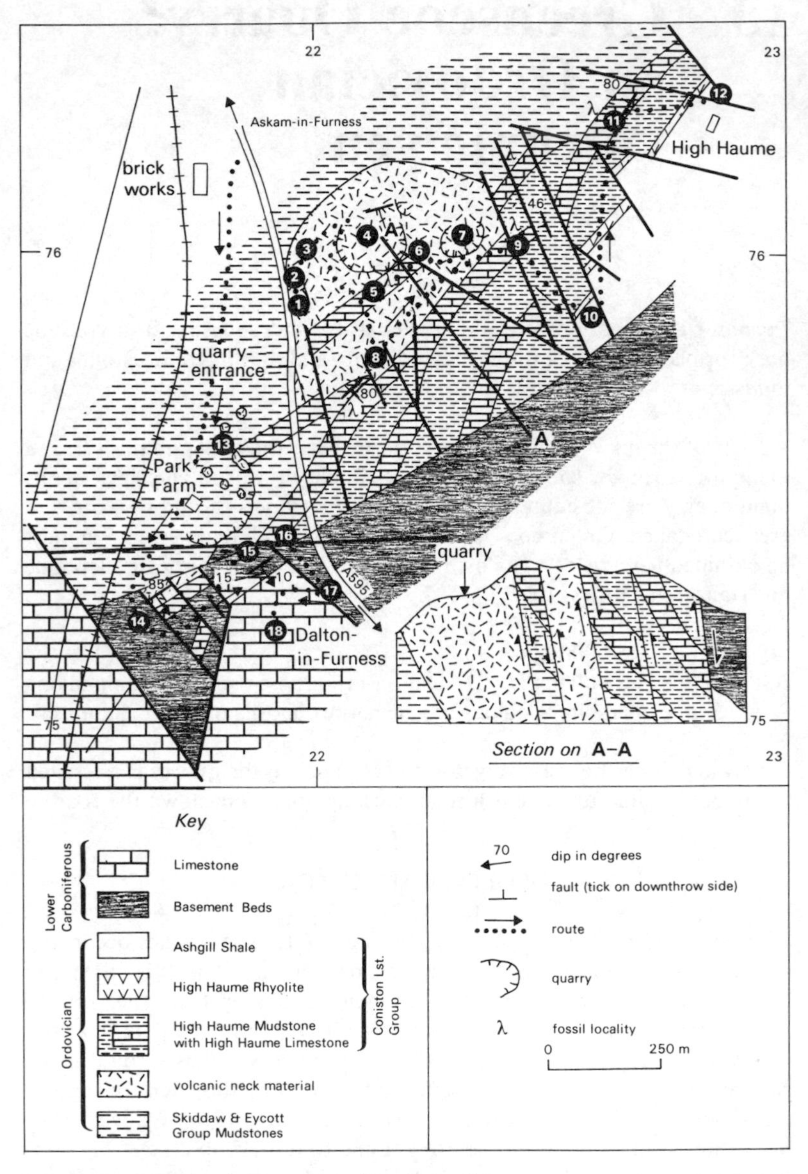

Figure 51 Geological map and section of the Greenscoe area.

Borrowdale Volcanic Group age. A second unconformity is present, with Lower Carboniferous Basement Beds resting on folded and eroded Lower Palaeozoic rocks (Fig. 51).

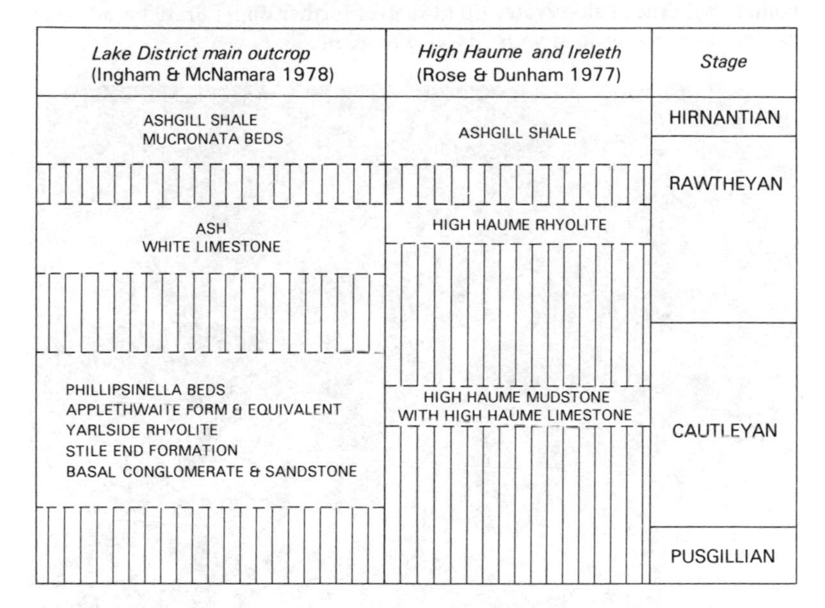

Figure 52 Nomenclature of Coniston Limestone Group (Rose & Dunham 1977). Modified to show non-sequence at base of Ashgill Shale.

The interrelationships between the lithologies seen in a volcanic neck are complex since they probably result from more than one phase of volcanic activity. However, certain features such as inclusions of country rock, lava sheets and various grades of **agglomerate** are typical of volcanic necks and can easily be identified. This itinerary has been designed to demonstrate these features.

EXCURSION DETAILS

From the entrance of the quarry (220758) proceed 200 m passing small exposures of fine agglomerate on the right. Location **1** is at the junction of a side track where a small quarry on the right is hidden by trees. At the northern side can be seen a 3 m thick sheet of andesite making a contact of about 53° with fine agglomerate.

Returning to the track, about 20 m further on a grass track leads off to the right and here at **2** a fine **tuff** can be seen, followed after 30 m by more andesite overlain by an agglomerate with slickensiding trending 328°. This is followed in the next 30 m by more andesites which make a vertical contact with the agglomerate. Eight metres further on indurated mudstone can be seen interbedded with fine agglomerate.

Figure 53 Greenscoe Quarry, locality **4**. Coarse agglomerate containing sub-rounded clasts of tuff (light) and andesite (dark). (Photograph: D. Leviston.)

Location **3** (220760) is at the old chute just past a track leading to the left. As you face the chute, to the left can be seen indurated mudstones which give way on the right-hand side to a tuffaceous rock. Location **4** is in the main quarry which consists of coarse agglomerate (Fig. 53).

Rose and Dunham (1977) stated that the main quarry appears to be sited in the central part of the vent complex, which consists of a roughly circular outcrop of coarse agglomerate with associated fine-grained agglomerate, tuff and small intrusions of andesite. Both tuffs and agglomerates appear to have formed in the vent from material pulverised by the passage of hot volcanic gas under fluidising conditions. The agglomerates contain sub-angular to sub-rounded pyroclasts of andesite, tuff and country rock (bluish-grey mudstone) in a matrix of tuff and pulverised mudstone. The pyroclasts vary in size up to 3 m long in the case of the igneous rock fragments, whilst the occasional mudstone fragments are much smaller. Both tuffs and agglomerates are unbedded and unsorted, but they do show cleavage. The more massive andesites may be identified by their brown weathered appearance and absence of cleavage.

Now take the track leading southwestwards from the entrance to the main quarry. After 200 m bear left at the junction, **5** being 100 m up the track where a much-jointed limestone is exposed on the corner. This is the High Haume Limestone belonging to the Coniston Limestone Group (Fig. 52) and here it has been faulted into the volcanic neck. Follow the strike of the limestone, passing over a heavily dolomitised section readily identified by its brown weathering, and at **6** the faulted contact between the limestone and the agglomerate can plainly be seen. Silicified brachiopods have also been found. The High Haume Mudstones sequence is repeated southeastwards by a set of reverse and normal faults (Fig. 54). To the north-east, twin bands of the High Haume Limestone form prominent features, the western band being faulted against shales of Eycott age.

Figure 54 Greenscoe Quarry. View looking north-east from locality **6** showing neck agglomerate in the foreground with fault-bounded blocks of limestone to the right. Locality **10** is on the extreme right of the picture. (Photograph: D. Leviston.)

Continue along the top of the quarry to **7** where a fine-grained blue-grey crystal-lithic tuff containing brown inclusions of what may be goethite is exposed in the path whilst medium- and fine-grained crystal-**lithic tuffs** and andesitic **lapilli tuffs,** commonly flow banded, occur in the face separating the two quarries. At the bottom of the quarry an altered purplish-green amygdaloidal andesite, probably representing a late stage intrusion, or possibly a small subsided block of lava, forms a prominent feature.

Return now to the gate near **5**. Almost opposite to the east a small dark green intrusive porphyritic andesite can be seen faulted against almost vertical limestones at **8**.

At this point the excursion can be shortened by returning to the car park then to **13**.

Follow the depression, mapped by Rose and Dunham (1977) as being in the agglomerate, northwards. Notice how the topography is interrupted by the NW–SE-trending dip faults displacing the strike of the limestones. Pass through the gate at the top and to the east is **9** (224760). Here is a tilted block of coarsely mottled limestone, the top surface of which appears to have been reworked after lithification to form a rubbly bed accentuated by weathering. Rose and Dunham (1977) record a fauna of anthozoa, bryozoa, brachiopoda, mollusca and ostracoda from the quarry on the eastern side of this locality. Location **10** is High Haume Beacon (227160) where the High Haume Rhyolite is exposed with small exposures of Ashgill Shale nearby. The weathering of the rhyolite has caused it to be the subject of much discussion and it has on many occasions been mistaken for a quartzite. It can be traced northwards and is seen to be displaced by more NW–SE-trending faults. It has been given the same age as the tuff band seen at Appletreeworth (Excursion 12) and seems to represent local Ashgillian volcanicity.

An old quarry 200 m west of the High Haume (**11**) (228763) is in strongly cleaved mudstones that dip 80° to the NNW, which is opposite to the general dip of the limestones we have seen. This may indicate that these beds are overturned, which makes their relationship to the High Haume Limestone uncertain. A varied but poorly preserved fauna of trilobites, brachiopods and corals suggest they may be of Cautleyan age (Fig. 52). The mudstones are partly decalcified, weather brown and contain large calcareous concretions.

The only exposure of the Mucronata Beds is seen in the access road to High Haume Farm (**12**) where about 9 m of greyish green, brown-weathering cleaved mudstones are exposed. The locality lies within a fault belt and the mudstones are crushed and stained with haematite; however, fossils have been collected from these beds.

Now return to the car park, by following the gap between Greenscoe Quarry and the Limestone bands down to the A595.

From the car park follow the track to Park Farm passing exposures of mudstone of the Skiddaw Group. Location **13** is the shale pit where agglomerate can be seen intruded into the Skiddaw slates. Permission to enter these pits must be obtained from Park Farm. High up on the northern side High Haume Limestone can be seen resting upon agglomerate whilst just below massive andesite appears to be intruded horizontally into the agglomerate. More agglomerate occupies the southern side of the quarry where at one time a circular column of andesite about 6 m high and 3 m in diameter could be seen intruded into the Skiddaw Slates, but all that now remains is a pile of broken rock.

Proceed southwards to Park Cottage (**14**) (217753) where red conglomerates, shales and mudstones of the Lower Carboniferous Basement Beds can be seen exposed in the bank on the left-hand side of the track. Behind the cottage High Haume Limestone overlying volcanics can be seen with another fault-bounded inlier to the south. Location **15** is in the woods immediately behind Park Farm where more conglomerates and red siltstones are faulted against High Haume Limestone although the contact is not seen. At one time the unconformity between the Skiddaw Slates and the Carboniferous Basement Beds could plainly be seen in the shale quarry immediately behind Park Farm; but, alas, this excellent section has been destroyed.

Location **16** is further upstream in a cutting that was the old mineral railway line that took iron ore from the nearby mines down to the iron works at Askam-in-Furness. Here well bedded High Haume Limestone with a southeasterly dip can be seen. Follow the old mineral line in a southerly direction. The next outcrop of rocks at **17** is of Lower Carboniferous Limestones of early Chadian age. The rocks are a sequence of unfossiliferous dolomites, decalcified limestones and calcite mudstones. Twenty-eight metres to the south from the footpath a vertical wall of fault breccia indicates that the old mineral line has been excavated along a fault plane. A visit to the summit of Housethwait Hill (**18**) (219752) provides a good vantage point for viewing the area (Fig. 55).

After completing the itinerary no doubt you will agree that this area provides some exciting and diversified geology. Return to the car park via the A595.

Figure 55 Greenscoe Quarry, locality **18**. General view showing Greenscoe Quarry on the left with the Coniston Limestone Group forming the three hillocks on the skyline. (Photograph: D. Leviston.)

Appendix I Bibliography, references and excursion reports

EXCURSION

Arthurton, R. S., I. C. Burgess and D. W. Holliday 1978. Permian and Triassic. In *The geology of the Lake District*, F. Moseley (ed.), Ch. 13. Yorkshire Geol Soc. Occasional Publication no. 3. — 2,10

Bonner, I. H. (ed.) 1981. *Field studies in Cumbria*. Cumbria Education Committee, Carlisle.

Bott, M. H. P. 1978. Deep structure. In *The geology of the Lake District*, F. Moseley (ed.), Ch. 3. Yorkshire Geol Soc. Occasional Publication no. 3. — 1, 5, 11

Burgess, I. C. and A. J. Wadge 1974. *The geology of the Cross Fell area*. HMSO, London. — 2

Capewell, J. G. 1954. The basic intrusions and an associated vent near Little Mell Fell, Cumberland. *Trans Leeds Geol Soc.* **6**, 243–8. — 3

Capewell, J. G. 1955. The post-Silurian, pre-marine Carboniferous sedimentary rocks of the eastern side of the English Lake District. *J. Geol Soc. Lond.* **111**, 23–46. — 3

Davies, W. J. K. 1968. *The Ravenglass and Eskdale Railway*. Newton Abbot: David and Charles. — 11

Dwerryhouse, A. R. 1909. Intrusive rocks in the neighbourhood of Eskdale. *J. Geol Soc. Lond.* **65**, 55–80. — 11

Eastwood, T., E. E. L. Dixon, S. E. Hollingworth and B. Smith 1931. The Geology of the Whitehaven and Workington District. *Mem. Geol Survey*. — 10

Eastwood, T., S. E. Hollingworth, W. C. C. Rose and F. M. Trotter 1968. *Geology of the country around Cockermouth and Caldbeck*. Mem. Geol Survey. — 6, 7

Firman, R. J. 1978. Intrusions. In *The geology of the Lake District*, F. Moseley (ed.), Ch. 10. Yorkshire Geol Soc. Occasional Publication no. 3. — 1, 5, 6, 11

Green, J. F. N. 1918. The Mell Fell Conglomerate. *Proc. Geol Assoc.* **29**, 117–25. — 3

Green, J. F. N. 1918. The Skiddaw Granite. A structural study. *Proc. Geol Assoc.* **29**, 126–36. — 5

Hitchen, C. S. 1934. The Skiddaw Granite and its residual products. *Q. J. Geol Soc. Lond.* **90**, 158–200. — 5

Ingham, J. K. and K. J. McNamara 1978. The Coniston Limestone Group. In *The geology of the Lake District*, F. Moseley (ed.), Ch. 9. Yorkshire Geol Soc. Occasional Publication no. 3. — 12

Jackson, D. E. 1978. The Skiddaw Group. In *The geology of the Lake*

EXCURSION

District, F. Moseley (ed.), Ch. 7. Yorkshire Geol Soc. Occasional Publication no. 3. — 8, 9

Jones, T. A. 1915. On the presence of Tourmaline in Eskdale (Cumberland) Granite. *Geol. Mag.* **2**, 190. — 11

Marr, J. E. 1916. *The geology of the Lake District*. Cambridge: Cambridge University Press. — 5

Mitchell, G. H. 1956. The Borrowdale Volcanic Series of the Dunnerdale Fells, Lancashire. *Liverpool Manchester Geol J.* **1**, 428–49. — 12

Moseley, F. 1964. The succession and structure of the Borrowdale volcanic rocks north-west of Ullswater. *Liverpool Manchester Geol. J.* **4**, 127–42. — 3

Moseley, F. 1972. A tectonic history of North-West England. *J. Geol Soc. Lond.* **128**, 561–98. — 3

Moseley, F. 1973. Structural relations between the Skiddaw Slates and the Borrowdale Volcanics. *Proc. Cumberland Geol Soc.* **3**, 127–45. — 8, 9

Moseley, F. 1977. Caledonian plate tectonics and the place of the English Lake District. *Geol Soc. Am. Bull.* **88**, 764–8.

Moseley, F. 1981. *Methods in field geology*. San Francisco: W. H. Freeman.

Moseley, F. (ed.) 1978. *The geology of the Lake District*. Yorkshire Geol Soc. Occasional Publication no. 3.

Rastall, R. H. 1910. The Skiddaw Granite and its metamorphism. *Q. J. Geol Soc. Lond.* **66**, 116–40. — 5

Rose, W. C. C. and K. C. Dunham 1977. Geology and haematite deposits of South Cumbria. *Econ. Mem. Geol Survey* Sheet 58, part 48. — 12, 13

Shackleton, E. H. 1968. *Lakeland geology*. Clapham: Dalesman. — 1, 5, 6, 8

Shackleton, E. H. 1975. *Geological excursions in Lakeland*. Clapham: Dalesman Publication Co. — 4, 12, 13

Simpson, B. 1934. The petrology of the Eskdale (Cumberland) Granite. *Proc. Geol. Assoc.* **45**, 17–34. — 11

Skillen, I. E. 1973. The igneous complex of Carrock Fell. *Proc. Cumberland Geol Soc.* **3**, 71–85. — 6

Smith, R. A. 1967. The deglaciation of South-West Cumberland. A re-appraisal of some features in the Eskdale and Bootle areas. *Proc. Cumberland Geol Soc.* **2**, 76–83. — 11

Smith, R. A. 1974. *A bibliography of the geology and geomorphology of Cumbria*. Workington: Cumberland Geol. Soc.

Trotter, F. M., S. E. Hollingworth, T. Eastwood and W. C. C. Rose 1937. *The Geology of the Gosforth District*. Mem. Geol Survey. — 11

Wadge, A. J. 1978. Classification and stratigraphical relationships of the Lower Ordovician rocks. In *The Geology of the Lake District*, F. Moseley (ed.), Ch. 6. Yorkshire Geol Soc. Occasional Publication no. 3. — 12, 13

EXCURSION

Wadge, A. J. 1978. Devonian. In *The geology of the Lake District*, F. Moseley (ed.), Ch. 11. Yorkshire Geol Soc. Occasional Publication no. 3. 3

Wainwright, A. 1978. *Walks from the Ratty*. Ravenglass: The Ravenglass and Eskdale Railway Co. 11

Ward, J. C. 1876. *The geology of the northern part of the English Lake District*. Mem. Geol Surv. 4

Wilson, P. 1975. The Rosthwaite Moraines. *Proc. Cumberland Geol. Soc.* **3**, 239–49. 8

The following excursion reports have appeared in the Proceedings of the Cumberland Geol Soc. (PCGS):

Buckley, D. K. 1969. The Greenscoe area. *PCGS* **2**, 178–81. 13

Burton, M. F. 1967. Devoke Water. *PCGS* **2**, 98–100. 11

Burton, M. F. 1969. Armboth Dyke. *PCGS* **2**, 175–7. 4

Burton, M. F. 1974. Intrusive rocks in the Eskdale area. *PCGS* **3**, 168–9. 11

Burton, M. F. 1977. Armboth Dyke. *PCGS* **4**, 59–61. 4

Dockwray, F. E. 1963. Carrock Fell Gabbro. *PCGS* **1** (1963), no. 2, 19–22. 6

Shackleton, E. H. 1968. Appletreeworth Beck. *PCGS* **2**, 117–18. 12

Shackleton, E. H. 1977. Carboniferous limestones of the area around Caldbeck and Uldale. *PCGS* **4**, 46–50. 7

Shilston, A. M. and J. R. Harpum 1964. The Shap Granite area. *PCGS* **1** (1964), no. 1, 32–6. 1

Shipp, T. 1975. The Skiddaw Granite in Sinen Gill. *PCGS* **3**, 255–6. 5

Smith, R. A. 1968. Deglaciation of South-West Cumberland. *PCGS* **2**, 119–21. 11

Smith, R. A. 1976. The Corney and Bootle areas of South-West Cumbria. *PCGS* **3**, 260–4. 11

Publications of the Cumberland Geological Society may be obtained from the Publications Secretary: Mr E. Lawrence, 2 Vicarage Lane, Cockermouth, Cumbria CA13 9DG.

General enquiries to the Cumberland Geological Society should be addressed to the General Secretary: Mr T. Shipp, Crook Hall, 28 High Seaton, Workington, Cumbria CA14 1PD.

Appendix II Museum facilities in Cumbria

The Lake District National Park Centre, Brockhole, Windermere. Telephone Windermere (096 62) 2231. There are excellent facilities for visiting individuals or groups. There is a comprehensive and well laid out display of geological material relating to the Lake District. There are good audio-visual facilities, with lectures and films arranged by request in advance of a visit. A charge is made for entrance to the Centre. The Centre also maintains the Lake District records of the National Scheme for Geological Site Documentation.

Tullie House Museum and Art Gallery, Carlisle. Telephone Carlisle (0228) 34781. This has a good exhibition of Cumbrian geological material.

Abbott Hall Art Gallery and Museum, Kendal. Telephone Kendal (0539) 22464. This has an exhibition of Lakeland life and industry, with geological exhibits related to the industries (principally mining and quarrying) and end-products of the region. Admission 15p (students free on production of student card).

Kendal Museum of Natural History and Archaeology, Station Road, Kendal. Telephone Kendal (0539) 21374. This has a recently revitalised display of the natural history and geology of the Lake District. Admission 10p (students free on production of student card).

Fitz Park Museum, Station Road, Keswick. Telephone Keswick (0596) 73263. Contains a heterogeneous collection of local memorabilia. There is some local geological material. Satisfactory for a casual visit on a wet afternoon.

Whinlatter Forest Visitor Centre, Braithwaite. Situated near the summit of Whinlatter Pass. Telephone Braithwaite (059 682) 469. This holds a very small geological display, mostly graptolites, loaned by E. H. Shackleton. The centre holds a well planned exhibition of Forestry Commission achievements and intentions. There are good audio-visual facilities with lectures and films (on forestry) arranged by request in advance of visit.

Whitehaven Museum, Market Place, Whitehaven. Telephone Whitehaven (0946) 3111, ext. 289. Contains displays on local geology, coal, fireclay and iron ore mining, and local pottery.

Ruskin Museum, The Institute, Yewdale Road, Coniston. Telephone Coniston (096 64) 359. Contains John Ruskin's memorabilia and includes his mineral collection. Admission 10p.

Grizedale Forest Visitor and Wildlife Centre, Grizedale, Hawkshead, Ambleside. Telephone Satterthwaite (022 984) 273. Contains exhibits of forest industries, wildlife, geology and industrial archaeology. Admission 10p.

Dalehead Base, Seatoller, Keswick. Telephone Borrowdale (059 684) 294. Contains a study area, useful on wet days. Admission free.

Appendix III Glossary

This Glossary is by no means exhaustive. The names of the more common rocks and minerals, and general geological terminology, have been assumed to be understood by the reader. In addition, items adequately explained in the text do not appear in the Glossary.

agglomerate A volcanic rock formed mainly of blocks or fragments generally more than 50 mm in diameter and ejected in a plastic state from a volcano.

amphiboles A group of ferromagnesian silicates occurring widely in igneous and metamorphic rocks. Hornblende is the commonest of this group of minerals.

amygdales Gas bubbles in lava, usually elongated and flattened somewhat by having been drawn out in the direction of flow of lava, while molten, to form almond-shaped cavities which subsequently became infilled with secondary minerals such as quartz, calcite or chlorite.

andalusite Aluminium silicate (Al_2SiO_5) resulting from low-grade thermal metamorphism of shales or mudstones. It forms elongated, prismatic crystals which, if they contain dark cross-shaped inclusions, are known as **chiastolite**.

andesite A fine-grained igneous rock, dark grey, green or brown in colour with **phenocrysts** of white **feldspar, biotite** or **hornblende**, sometimes **augite**.

anhydrite A mineral; a colourless or white, sometimes tinted form of calcium sulphate, often associated with gypsum but distinguished from it by its hardness, 3–3½ on Mohs' scale, and by the absence of water of hydration.

aplite A light-coloured, fine-grained igneous rock found in thin veins penetrating granite. It probably results from the final stages of consolidation of a body of magma.

augite See **pyroxene**.

azurite A deep azure-blue mineral, copper carbonate, often associated with **malachite**.

barite (barytes) A white or colourless, sometimes tinted, mineral, distinguished from quartz and calcite by its high relative density (4.5).

batholith A large intrusive mass of granitic rock associated with, and aligned along, an orogenic belt. The irregular upper surface of a batholith frequently forms a series of discontinuous outcrops of granite such as is seen in the Lake District. Each outcrop has had a heating effect on the rocks it has intruded, giving rise to a **metamorphic aureole**.

biotite A brown, black or dark-green mica.

boss A cylindrical mass of plutonic igneous rock whose diameter is typically in excess of 1 km. The Shap Granite occurs in the form of a boss.

Brockram Breccia of Permo-Triassic age found in the Vale of Eden and in West Cumbria.

BVG Borrowdale Volcanic Group.

Caledonian A period of **orogeny** caused by a collision of continents across northern Britain and Scandinavia around 400 Ma ago.

cerussite A white or grey mineral, lead carbonate, with a very high relative density (6.5).

chalcopyrite (copper pyrites) A copper mineral, brassy yellow in colour, often with iridescent tarnish; resembles iron pyrites (pyrite) but is softer, 3½–4 on Mohs' scale.

chiastolite See **andalusite**. Chiastolite 'slate' has resulted from the low-grade thermal metamorphism of mudstones around the periphery of the Skiddaw Granite.

chlorite A group of green, complex iron–magnesium–aluminium silicates resulting from the alteration of such minerals as hornblende and biotite. The generally greenish tinge of many rocks of the Borrowdale Volcanic Group is due to the presence of chlorite.

clast A fragment of pre-existing rock, for example a pebble in a conglomerate. A **phenoclast** is a clast that is markedly larger than the fragments that surround it.

cleavage The development of planes of splitting in rocks, usually parallel to the axial planes of folding, and thus independent of bedding. Rocks showing good cleavage (originally shales, mudstones or tuffs) have been subjected to regional metamorphic pressures resulting in the realignment and/or recrystallisation of flaky minerals, and are generally termed **slates**.

cobble A rock fragment, usually rounded to some extent, varying in diameter from 64 mm to 256 mm: i.e. between a pebble and boulder in size.

copper pyrites See **chalcopyrite**.

cordierite An iron–magnesium cyclosilicate commonly found as a constituent of **hornfels** resulting from the thermal metamorphism of clay rocks.

cyclothem A cycle of sedimentation. See p. 9.

dacite A volcanic lava of composition some way between that of rhyolite and **andesite**.

diabase An altered **dolerite**.

diorite A coarse-grained igneous rock containing white **feldspar** and dark **hornblende** or **biotite**, sometimes **augite**, giving the rock a speckled appearance.

dolerite A medium-grained igneous rock containing **plagioclase** and **pyroxene (augite)**. When fresh it is black or dark green.

dolomite A white, yellow or brown mineral, the double carbonate of calcium and magnesium; reacts very slowly with dilute hydrochloric acid, unlike calcite, and is harder than that mineral (3½–4 on Mohs' scale). The name also is applied to a rock composed of the mineral.

dolomitisation The complete or partial conversion of limestone ($CaCO_3$) into **dolomite** rock ($CaMg(CO_3)_2$) by the interchange of Ca^{2+} and Mg^{2+} ions during the percolation of highly saline water through the rock.

dyke A wall-like body of igneous rock cutting across the bedding of the host rock.

epidote A group of pale yellowish-green silicate minerals, commonly found in rocks of the Borrowdale Volcanic Group, resulting from low-grade regional **metamorphism**, or from **metasomatism** adjacent to an igneous intrusion.

erratic A boulder that has been removed some distance from its source at outcrop by the effect of glacial transport.

euhedral A fully developed crystal form.

evaporite A mineral precipitated by the evaporation of saline water. Typical evaporite minerals are halite ($NaCl$), gypsum ($CaSO_4.2H_2O$) and **anhydrite** ($CaSO_4$).

felsite A fine-grained intermediate to acid igneous rock generally found as **dykes**.

feldspar (felspar) The most important of the rock-forming minerals, found in the great majority of igneous rocks. Variations in constitutions of magma and conditions of crystallisation give rise to a variety of feldspars of which the most important are **orthoclase** ($KAlSi_3O_8$), and the **plagioclase** series formed of varying proportions of albite ($NaAlSi_3O_8$) and anorthite ($CaAl_2Si_2O_8$).

flow-banding Alignment of crystallising particles within a flowing magma or lava. Sometimes called fluxion banding.

fluorite (fluorspar) A colourless, white, yellow, green or purple mineral, calcium fluoride, hardness 4 on Mohs' scale; may be found as cubic crystals.

flute cast A groove eroded on a muddy sea bed by a turbidity current and subsequently infilled with coarser sediment.

gabbro A coarse-grained igneous rock consisting essentially of **plagioclase** and **pyroxene (augite)**. Its colour and appearance depend partly on that of the plagioclase, which is white to dark green or grey.

garnet Complex silicates of varying composition and colour, forming rounded crystals and found most commonly in metamorphic rocks. They are distinguished by their form and hardness (6–7½ on Mohs' scale). Almadine is a common red variety.

greisen A quartz- and mica-rich rock produced by the pneumatolysis of granite by fluorine-rich vapours.

greywacke A sandstone or siltstone composed mainly of angular to sub-angular particles that are mainly rock fragments (rather than quartz). They are generally poorly sorted, and often show evidence of very rapid deposition, for example, from turbidity currents.

haematite (hematite) Iron oxide and ore. Steel grey to black when pure, red when impure, with a brick-red powder or streak. It may take rounded forms resembling a kidney or dish of kidneys **(kidney iron ore)**.

Hercynian A period of orogeny in the late Palaeozoic mainly affecting regions of northwestern France and southwestern Britain. Sometimes called the **Variscan** or Armorican orogeny.

hornblende A mineral of the amphibole group.

hornfels A hard, fine-grained rock produced by intense thermal **metamorphism** of clay rocks. Commonly contains the minerals **andalusite** and **cordierite** in random orientation.

ignimbrite A rock formed by a hot volcanic ash flow. The fragments may become welded together while still hot to form a welded tuff.

ilmenite A black oxide of iron and titanium usually occurring as microscopic crystals in igneous rocks such as gabbro.

joints A pattern of cracks in a sheet of rock across which there has been no lateral movements. Joint patterns may form by shrinkage (as in cooling lava), decompression (by rapid removal of overburden), or flexuring of rock during folding. Joints formed at the crests of folds are particularly prone to infilling with secondary minerals such as quartz and are then termed **tension gashes**.

kaolin China clay, one of the main products of chemical decomposition of **feldspars** by weathering or hydrothermal action.

kidney ore See **haematite**.

lapilli tuff A pyroclastic rock formed from an accumulation of small, peanut-sized pieces of volcanic rock blown out during an eruption.

lithic tuff Almost synonymous with **lapilli tuff**, having formed from small pieces of rock (as opposed to crystals such as **augite**).

load cast A bulbous impression on the base of a bed of sandstone formed where it was forced under pressure into an underlying layer of mud before consolidation took place.

Ma Symbol for million years.

malachite A bright green copper carbonate mineral, often found as an encrustation in association with **azurite**.

meltwater channel Valley eroded by water torrents from melting ice. Such channels may have formed at the base of an ice sheet with water moving, under considerable hydrostatic pressure, up and down over obstructions. On the disappearance of the ice, the channels remain as anomalous features of the landscape.

metamorphism Changes brought about in rock by heat, pressure and chemically active fluids. **Thermal** metamorphism involves mainly heat and is normally associated with igneous intrusion, the hot intrusive mass developing a **metamorphic aureole** by heat flow into the surrounding rocks. **Regional** metamorphism is associated with the compressive forces of **orogeny** and results in the formation of many new minerals and the widespread development of slaty **cleavage** in suitable rocks.

metasomatism A metamorphic change brought about by chemically active fluids migrating through a rock.

millet-seed grains Quartz grains showing perfect rounding by wind abrasion. Sandstones, such as the Penrith Sandstone, containing grains of this type are thought to have originated in a wind-swept desert.

molybdenite Molybdenum sulphide (MoS_2), a very soft silvery mineral.

oligoclase A variety of **plagioclase feldspar**.

olivine basalt A basalt containing original olivine, a yellowish-green clear mineral of hardness 6½–7 on Mohs' scale. This mineral readily alters in surface rocks.

orogeny A period of mountain building. The process of orogenesis leads, over a period of several tens of millions of years, to the intense deformation of a geosynclinal basin, with widespread **metamorphism** and granite emplacement.

orthoclase The commonest of the potassium **feldspars** ($KAlSi_3O_8$).

orthoquartzite A sandstone with silica cement.

pelitic A term used for metamorphic rocks derived from clays, shales and other argillaceous sedimentary rocks.

pencil slate A form of slate which has developed two sets of **cleavage** planes almost at right angles, with the result that it weathers into stick-like, rather than platy, fragments.

perthite An intergrowth of **orthoclase** and albite **feldspars**.

phenoclast See **clast.**

phenocryst A relatively large crystal set in a finer groundmass in an igneous rock. The texture is known as **porphyritic**.

plagioclase A series of sodium and calcium feldspars of varying proportions.

porphyritic texture See **phenocryst**.

porphyroblast A relatively large crystal that has grown in a finer-textured metamorphic groundmass during the process of **metamorphism**.

pyromorphite A green, yellow or brown mineral, chlorophosphate of lead, with a high relative density (about 7).

pyroxene A group of ferromagnesian minerals occurring widely in igneous rocks. **Augite** is the commonest member.

pyrrhotite A coppery-bronze iron sulphide mineral often found in basic igneous rocks.

ripple drift lamination Lamination produced by current-formed ripples during deposition of sediments.

roche moutonnée A glacial erosional feature consisting of an asymmetrical lump of bedrock with a smoothed, striated face upstream and a plucked, craggy face downstream to the direction of ice flow.

sericite A type of mica commonly found in **greisen**.

slates Clay rocks that have been subjected to low-grade regional metamorphism that has brought about a well defined **cleavage**. Slates subjected to thermal metamorphism may develop spotting, **chiastolite** crystals and **cordierite** crystals with increasing thermal stress.

slickensiding Polishing and grooving developed along a fault plane by the abrasive action of rock sliding over rock.

sole markings Irregularities on the undersurface of a layer of rock resulting from the manner in which it was deposited. Sole markings include **flute casts** and **load casts.**

sphalerite See **zinc blende**.

stock A large intrusion, usually granitic, having steeply dipping sides and no apparent floor.

stope (a) Extraction of veinstone from above a tunnel in mining; (b) the upward movement of magma along joints and fissures. The existence of broken-off pieces of country rock (**xenoliths**) within igneous intrusions lends evidence to the process of magmatic stoping.

stringers Mineral veinlets, often of quartz.

tension gash See **joints.**

till A mixture of assorted rocks and clay of glacial origin: boulder clay.

tourmaline A complex silicate mineral resulting from pneumatolysis in the vicinity of some granites. The black variety is known as schorl.

tufa A deposit of calcium carbonate formed by deposition from groundwater in limestone districts. It may fill joints or form a spongy deposit around plants, etc.

tuff Consolidated volcanic ash.

turbidite A rock unit formed by the settling of material from suspension in water following the passage of a turbidity current. The latter is a slurry of mud, sand and water flowing, often at high speed, down an underwater slope, and travelling on across the flat sea floor for a considerable distance before settling out.

twinning Arrangement of the crystal lattice in such a way that one part is a reflection of the other. For instance **orthoclase feldspar** shows a single twin plane which may be easily demonstrated by angling the light off the feldspar crystals in a piece of Shap Granite. **Plagioclase feldspar** crystals show many closely spaced parallel (lamellar) twin planes which are best seen by looking at a thin slice in polarised light through a microscope.

unconformity A rock surface upon which rests a layer of rock formed very much later. The older rocks may have been tilted and eroded before the newer rocks were laid over them.

Variscan See **Hercynian**.

vesicular texture Produced by bubbles of gas in volcanic lavas. Sometimes elongated in direction of flow. **Vesicles** are often infilled with secondary minerals. See **amygdales**.

wadi A water run-off channel or valley in a desert or semi-desert region. Wadis usually only contain water intermittently.

welded tuff See **ignimbrite**.

wollastonite Calcium silicate mineral ($CaSiO_3$) formed by thermal **metamorphism** of impure limestone.

wrench fault A vertical fracture where the rocks on either side have slid horizontally relative to each other.

xenolith An inclusion of country rock in an igneous intrusive rock. See **stope**.

zinc blende (sphalerite) A mineral, zinc sulphide, light to dark brown or black; hardness 4 on Mohs' scale.

Index

Page numbers in *italics* refer to text figures.

adit 46, 76, *107*
agate 91
agglomerate 7, 36, 115, 117–20, *118–19*, 126
albite 128, 130
almandine 52, 128
amphibole 19, 126, 128
amygdale 19–20, 111, 126, 131
andalusite 41, 54–5, 126, 128
andesite 19, 72, 75, 83, 91, 95–102, 103–5, 107–11, 115, 117–20, 126–7
anhydrite 26, 86–90, 126, 128
ankerite 55–7
anorthite 128
anticlinal axis 10, *33*
anticline 8, 17
apatite 56
aplite 126
Appleby 21, 24–5
Appletreeworth 107–14, *107–12*, 120
arête 76, 83
Armathwaite Dyke *23*, 26
Armboth Dyke 36–9, *36*
Armorican orogeny 128
arsenopyrite 55, 57
Ashgill Shales 112–13, *116–17*, 120
augite 71, 111, 126–7, 129–30
Aulophyllum 62–3
axial plane cleavage 45, 127
azurite 113, 126, 129

Bannisdale Slate 16
barite (barytes) 18, 46, 68, 113, 126
Barrowmouth 87, *88–9*, 89
basalt 31, 129
batholith 3, 8, 17, 41, 95, 126
Beckfoot 93, *101*, 102–3
bedding plane 60, 62, 77, 79, 91, 127
Belah Dolomite *22*, 25
biotite 41, 52, 99, 126–7
bismuthinite 56
bivalve 90
Bleaberry Combe 76
Blencathra 53, 68
Boot 93, 95, *100*, 102–3
Borrowdale 8, 65–73, *66–72*, 83
Borrowdale Volcanic Group (BVG) 3, *4–16*, 7, 18–19, 27, *28–33*, 31–2, 36, 38, 51, 65, 68–9, *70*, 73–5, *75*, 77, 83, 90–1, 93, *94*, 95, *98–100*, 99, 102, 107, 109, 117, 126–7
boss 17, 95, 126
boulder clay 91, 96, 130
Bowder Stone 72
Bowscale Tarn 50, 53–4, *53*
brachiopod 31, 62–3, 90, 111, 119–20
Brandy Gill 50, 55–7
Brathay Flags 114, *117*
breccia 24, 87, 109–10, *109–10*, 113, 126
Brockram *4*, 22, *22*, *23*, 24–5, 86–7, *88*, 90, 126
Brough 21, 25
Browgill Beds *107*, 114
Bryozoa 62, 120
Burrells 24–5
burrows *61*, 62, *63*
Buttermere 74–85, *75–8*, *82*
Buttermere and Ennerdale Granophyre 8, 95, 99

calcite 19, 29, 39, 56, 60, 72, 110, 121, 126–7
Caldbeck 50, 57–62
Caldbeck Fells 57
Caldew 40, 52–4
Caledonian orogeny *5*, 8–9, 79, 95, 126
campylite 57
Caninia 64
Carboniferous 3, *4–16*, 9, 20, 22, 24, 27, 30, 62, 85–7, *88*, 90–1, 121
Carboniferous Limestone *5–6*, 9, 24, 27, *28*, 30–1, 32, 53, 58–64, *59–63*
carnelian 91
Carrock Fell *43*, 50–8, *51*
Carrock Granophyre 51
Carrock Mine 8, 50, 55
Castle Crag 69–70, 83
Castle Head 7, 73
Cautleyan 120
cerussite 68, 127
Chadian 121
chalcopyrite (copper pyrites) 18, 46, 55, 113, 127

chiastolite 41, 44, 126, 130
China clay 129
chlorite 36, 72, 108, 126–7
cirque 10, 76
clast 24, 30–1, 109–10, *118*, 127, 130
cleavage 17, 45, *47*, 55, 70, 73, 77, 79, *80*, 111, 119, 127, 129–30
coal 9, 60, 64
Coal Measures *5–6*, 9, 86, *88*
cobble 29, 34, 127
combe 76–7
conglomerate 20, 27–34, 111, 121, 127
Coniston Grit *16*
Coniston Limestone Group *4–16*, 8, 19–20, 108–9, *109*, 111, *112*, 115, *117*, 119, *121*
copper 8, 44, 57, 76, 127
coral 31, 62–3, 90, 120
cordierite 41, 44, 54–5, 127–8, 130
corrie 50, 83
Cowrake Quarry 23
crinoid 62
cross bedding 24, 31, 91
Cross Fell Inlier 21
Crummock Water 74, 76–7
Cumberland Geological Society vii, viii, 124
  Proceedings 52, 64, 71, 123–4
Cumbria vii, viii, 9, 90
cyclothem 9, 127
Cyrtograptus 115

dacite 75, 127
delta 9, 30, 70
Derwent 65, 68–71
Derwentwater 65–8, 70, 73
desiccation crack 89
Devonian *4–5*, 9, 20, 29, 79
diabase 50, 54, 127
Dibunophyllum 63
Didymograptus 115
diorite 95, 99, 127
dip fault 107, 120
dolerite 26, 54, 73, 127
dolomite 25, 56, 121, 127
dolomitise 24, 90, 119, 127
dome 3, *6*, 10
drumlin 21, 73
dry valley 25
Drygill Shales 50, *51*, 57
dune bedding 24
Dunnerdale Tuffs 107
dyke 20, 26, 36, 38, 127

Eden Shales *5*, 22, 25
Eden Valley 9, 21–6, *23*, 126
epidote 19, 127
erratic 37, 91, 96, 127
escarpment 24, 27, 30, 109
Eskdale Granite 8, 93–105, *94–104*
euhedral 36, 128
evaporite *5*, 25–6, 87, 128
Eycott Volcanic Group *5–6*, 7, 51, *51*, 115, *116*

false bedding 90
Faulds Brow Quarry 58, 60, *60–3*
fault 10, 18, 24, 26, 32, 36, 38–9, 44, 55, 58, 76, 79, *81*, 83, 91, 95, 102–3, 107, 109, 112–13, 119–21, 130
feldspar (felspar) 38, 41, 45, 52, 55, 126–30
felsite 50, 51, 128
ferromagnesian 53, 126, 130
Fisher Gill 36, 37–8
Fleetwith Pike 75, 83
Fleswick 87, 91
flow banding 19, 38, 110–11, 119, 128
flow breccia 110
fluorite (fluorspar) 18, 56, 68, 128
flute cast 17, 128, 130
folding 8, *17*, *47*, 77, *77*, 79
foraminiferid 63
fossil 25, 57, 62–3, 90, 107, 112–13, 120

gabbro 50–3, 128
galena 46, 54–5, 68
garnet 19, 38, 52, 128
Gigantella 63
gilbertite 41, 56
glacial
  drift 72, 86, 88, 95, 111
  meltwater channel 25, 69, 71, 86–7, 93, 95–6
  moraine 65, 70–2, 87
  till 71, 87, 111
Glenderaterra Lead Mine 46, *48–9*
Goat Crag 7, *78*, 79, 81, 105
goethite 119
gossan 46, 57
graded bedding 79, *81*
Grainsgill 40, 50–1, 54
Grange 65, 68–9
granite 8, 15, 18–20, 40–5, 55, 93, 95–105, 126–31
granodiorite 41, 91, 95
granophyre 50–1, 53–4, 75

graptolite 54, 107, 113–15
gravitational anomaly 3, 41
Great Barrow 95, 103, *104*
Great Mell Fell 27, 31–2, 34
'green granite' 99, 103
Greenscoe 7, 115–21, *116*, *118–21*
greisen 41, 54, 56, 102, 128, 130
greywacke 17, 29, 41, 51, 91, 128
gypsum 26, 86–7, 90, 126, 128

haematite 9, 19, 23, 31, 93, 95, 102–3, 120
Harestones Felsite 50–1, *51*
Hassnesshow Beck 79, *78–80*, *81*
Hawk 107, *108*, 111, 113
Hercynian *5*, 52, 128
High Haume Limestone *116–17*, 119–21
High Haume Mudstone *116–17*, 119
High Haume Rhyolite *116–17*, 120
Honister 8, 75–6, 83, *84*
Hope Beck Slates 74, 77
hornblende 52, 99, 126–7
hornfels 20, 41, 44–5, 54, 127–8
hornfelsed slate 44, *46–7*, 52, 102

igneous intrusion *4*, 7, 18–19, 51, 99, 127, 129–30
ignimbrite 7, 111, 128
ilmenite 53, 128
inlier 3, 121
interturbidite 79
Irish Sea 9–10, 87, 95
iron 8, 18, 53, 76, 99, 111, 121, 128
Irton Pike 95, *98*, 99

Jaws of Borrowdale 70, *70*, 72
Jew Limestone *59*, 63
joint 17, 26, 31, 38, 45, 52, 60, 62, 77, 90–1, 102–3, 129, 131
joseite 56

kaolin(ised) 45, 129
Keswick 7, 50, 58, 65, 68, 73
'kidney ore' 103, 113, 128
Kirk Stile Slates 68, 74

Lake District vii, viii, 3–11, *4–6*, 21, 29, 41, 51–2, 58, 87, 91, 95, 126
landslip 89, 91–2
lapilli tuff 119, 129
lava 7, 31–3, 38, 68–9, 83, 90, 119, 126–9, 131
andesitic 7, 32–3, 72
basaltic 7
brecciated 110
flow-banded 38
rhyolitic 7
scoriaceous 83
lead 8, 44, 54, 57, 76
leucogabbro 53
Lickle Rhyolite 107, *112*
limestone 9, 19, 29–31, 58–64, 90, 109, 111–15, 119–20, 127, 131
Linbeck 95, *98*, 99
lithic tuff 119, 129
Lithostrotion 61, 62, 64
load cast 89–90, 129–30
Lonsdaleia 63
Lower Carboniferous 9, 31, 58–9, *116*, 117, 121
Lower Palaeozoic 3, 117
Loweswater Flags 74, 77, 79

magma 8, 18, 126, 128
Magnesian Limestone 86–7, *88*, 90
malachite 46, 126, 129
manganese 57, 113
marcasite 114
melagabbro 53
Mell Fell Conglomerate *4–5*, 9, 27–34, *28–33*
meltwater 10, 21, 87, 92
channel 65, 69, 71, 87, 93, 95, 129
metamorphic
aureole 7, 19, 41, *43*, 55, 95, 126, 129
zone 40–1, 44, 54
metamorphism 19, 38, 41–4, 74, 97, 126–30
metasomatism 15, 19, 41, 52, 127, 129
mica 54, 79, 130
biotite 41, 126
gilbertite 41, 56
muscovite 41
millet seed grain 24, 126
mineralisation 18–19, 44, 51–2, 55, 57–8, 74, 83, 110
minerals, list of properties 55–6
mineral vein 8, 19, 41, 50, 55, 57, 68, 130
Mohs' scale 55, 126–9, 131
molybdenite 18, 56, 129
Monograptus 114
moraine 65, 70–2, 87, 92
Mosedale 40, 50, 52, 54
Mucronata Beds 112, *117*, 120

mudflow 92
mudstone 8, 17, 41, 44, 51, 74–5, 83, 112, 114–15, 118–21, 126–7
museums 125
musical stones 44

New Red Sandstone 3, 21, 86–7
normal fault 113, 119

Old Red Sandstone 29
oligoclase 36, 41, 129
olivine basalt 31, 129
Ordovician vii, *4–5*, 7–8, *16*, 51, 57, 79, 107, *107*, 115, *116*
orogeny 126, 128–9
orthid 112
orthoclase 18, 20, 36, 41, 128–31
ostracoda 120

Palaeozoic 20, 128
pencil slate 33, 130
Pennine Fault 9, 25
Pennines 41, 53, 59
Penrith 7, 21–2, 24, 29, 31, 50
Penrith Sandstone *5*, 22–6, *22–3*, 129
perched block 37, 65
Permian *4–6*, 9, 21–2, *22–3*, 24, 26, 87, 90
Permo-Triassic 21, 24, *88*, 103
perthite 130
petrology 58
Phacops mucronatus 112
phenoclast 29, 127
phenocryst 18, *18*, 36, 38, 99, 111, 130
plagioclase 41, 52, 99, 128–31
Pleistocene vii, 91
pneumatolysis 128, 131
Pooley Bridge 27, 31–2
porphyritic texture 18, 99, 119, 130
porphyroblast 41, 44, 130
pressure solution 79
psilomelane 57
pyrite (iron pyrites) 18–20, 55, 60, 68
pyroclast(ic) 118, 129
pyrolusite 113
pyromorphite 46, 57
pyroxene 99, 111, 126–8, 130
pyrrhotite 38, 55, 130

quartz 18–19, 23, 36–8, 41, 46, 55–6, 68, 72, 79, *81*, 91, 99, 113, 126, 128–9
  gabbro 52–3
  porphyry 36
  sericite 102
  stringer 77
  vein 54–6, 68, 109
quartzite 24, 120
Quaternary *5*, 10
Quayfoot 65, 69, 71, 73, 83

radial drainage *6*, 7, 10
raised beach 11
Ravenglass 93, 96, *97*
Ravenglass and Eskdale Railway ('Ratty') 93, *94–7*, 99, 103
Red Brow 72, *72*
regional metamorphism 41, 127, 129–30
reverse fault 107, 109, 113, 119
rhyolite 19, 107, 112, 120, 127
ripple drift lamination 79, *81*, 130
ripple mark 31, 89, 91
roche moutonnée 65, 69–70, 72, *72*, 77, 95, 130
Rosthwaite 10, 65, *67*, 68, 70–2

Saccaminopsis 63
sandstone 7, 9, 21, 24–5, 31, 58, 60, 74, 89–91, 103, 111, 128–9
Scar Limestone *59*, 60, *62*
scheelite 56
Schidozus 90
schist 54–5
schorl 54, 131
scorodite 57
scree 10, 24, 64, 76, 87
sedimentary structures 9, 74, 86, 91
  basin 24
sericite 147
shale 9, 31, 57–62, 65, 87, 89–90, 109, 112, 115, 119–20, 126–7, 130
Shap Blue Rock Quarry 15, 19
Shap Granite 8–9, 15–20, *16*, *18*, 131
Shap Pink Rock Quarry 15
silica 23, 41, 95, 99, 129
siltstone 17, 20, 25, 31, 68, 121, 128
Silurian 3–9, *4–17*, 15, 17, 20, 41, 79, 107–8, *107*
Sinen Gill 40–1, *42*, 45–6, *45*
Skelgill Beds *107*, 109, 112–14
Skiddaw 3, 7, 41, 58, 68
Skiddaw–Borrowdale junction 65, *66*, 69, 83
Skiddaw Granite 7–8, 40–6, *42–5*, *51*, 52, 54–5, 127
Skiddaw Group 3–8, *4–16*, 27, *28–33*,

32, 40–1, 50–1, 53–5, 65, 68–9, 74–5, 77, *80*, 83, 115, *116*, 120
Skiddaw Slates 7, 32–3, 40–4, *43*, *46*, 52, 54, 65, 68–9, 73, 95, 109, 120–1
Skitwath Beck 30
slate 8, 44, 70, 76–7, 83, 127, 130
slickenside 24, 39, 44, *81*, 83, 112, 118, 130
sole marking 17, 77, 130
solifluction 76
sphalerite (zinc blende) 55, 130–1
spotted slate 54
St Bees Head 9, 86–92, *88*
St Bees Sandstone *5*, *22*, 86–91, *88*, 95
St Bees Shale *5*, *22*, 86–90, *88*
stock 41, 95, 130
submerged forest 92
swallow hole 63
syncline 17, 23, 107
Syringopora 61, 62

tension gash 17, 129–30
tetradymite 56
thermal metamorphism 19, 38, 40–4, 52, 74, 126–31
Thirlmere 36–8
thrust zone *33*, 79
till 71, 87, 111, 130
titanium 128
tombolo 77
tourmaline 54, 102, 131
trace fossil 62, *62*
trade winds 9
Triassic *4–6*, 7, 9–10, *23*, 91
trilobite 57, 62, 113, 120
trinuclids 57
trough end 76, 83
truncated spur 76, 83
tufa 31
tuff 7–8, 32–3, 36, 38, 51, 68–70, 73, 75, 83, 90–1, 107, 110–13, *110*, 115, 118–20, 127, 131
tungsten 8, 50
turbidite 74, 79, *80–1*, 131
turbidity current 79, 128, 131
twinning 36, 131

Ullswater 3, 10, 27–8, 32, *33*
Ullswater Anticline 32
Ullswater Thrust 32, 34
unconformity 15, 20, 86, 117, 121, 131
U-shaped valley 54, 95

Variscan orogeny 9, 22, 128, 131
vein 19, 44, 52, 55–7, 68, 91, 95, 126
vesicle 19, 109, 111, 131
volcanic
  breccia (see breccia)
  lava (see lava)
  neck 115–19
  plug 73
  tuff (see tuff)
  vent 7, 115, 118

Warnscale Crags 75, *82*, 83
waterfall 20, 45, *45*, 56–7, 76, 83, 103
'way up' of strata 81
weathering 3, 18
welded tuff 7, 111, 128, 131
Whitehaven 9, 86–7, 90
Whitehaven Sandstone 86, 90
wolfram 56
wollastonite 19, 131
wrench fault 36, 38, 107, 112, 131

xenolith 18, *18*, 20, 53, 95, 130–1

Zaphrentis 72
zinc 8
zinc blende (sphalerite) 68, 131
zone of metamorphism 41, 44, 54
Zoophycos 61–2, 62